AF806537

Unmanned System Technologies

Springer's Unmanned Systems Technologies (UST) book series publishes the latest developments in unmanned vehicles and platforms in a timely manner, with the highest of quality, and written and edited by leaders in the field. The aim is to provide an effective platform to global researchers in the field to exchange their research findings and ideas. The series covers all the main branches of unmanned systems and technologies, both theoretical and applied, including but not limited to:

- Unmanned aerial vehicles, unmanned ground vehicles and unmanned ships, and all unmanned systems related research in:
- Robotics Design
- Artificial Intelligence
- Guidance, Navigation and Control
- Signal Processing
- Circuit and Systems
- Mechatronics
- Big Data
- Intelligent Computing and Communication
- Advanced Materials and Engineering

The publication types of the series are monographs, professional books, graduate textbooks, and edited volumes.

More information about this series at http://www.springer.com/series/15608

Nishu Gupta • Arun Prakash • Rajeev Tripathi
Editors

Internet of Vehicles and its Applications in Autonomous Driving

Springer

Editors
Nishu Gupta
Department of Electronics
and Communication Engineering
Vaagdevi College of Engineering
Warangal, Telangana, India

Arun Prakash
Department of Electronics
and Communication Engineering
Motilal Nehru National Institute
of Technology Allahabad
Prayagraj, Uttar Pradesh, India

Rajeev Tripathi
Department of Electronics
and Communication Engineering
Motilal Nehru National Institute
of Technology Allahabad
Prayagraj, Uttar Pradesh, India

ISSN 2523-3734 ISSN 2523-3742 (electronic)
Unmanned System Technologies
ISBN 978-3-030-46334-2 ISBN 978-3-030-46335-9 (eBook)
https://doi.org/10.1007/978-3-030-46335-9

© Springer Nature Switzerland AG 2021
This work is subject to copyright. All rights are reserved by the Publisher, whether the whole or part of the material is concerned, specifically the rights of translation, reprinting, reuse of illustrations, recitation, broadcasting, reproduction on microfilms or in any other physical way, and transmission or information storage and retrieval, electronic adaptation, computer software, or by similar or dissimilar methodology now known or hereafter developed.
The use of general descriptive names, registered names, trademarks, service marks, etc. in this publication does not imply, even in the absence of a specific statement, that such names are exempt from the relevant protective laws and regulations and therefore free for general use.
The publisher, the authors, and the editors are safe to assume that the advice and information in this book are believed to be true and accurate at the date of publication. Neither the publisher nor the authors or the editors give a warranty, expressed or implied, with respect to the material contained herein or for any errors or omissions that may have been made. The publisher remains neutral with regard to jurisdictional claims in published maps and institutional affiliations.

This Springer imprint is published by the registered company Springer Nature Switzerland AG
The registered company address is: Gewerbestrasse 11, 6330 Cham, Switzerland

The editors dedicate this book to their Parents

Foreword

Internet of Vehicles and Its Applications in Autonomous Driving

I am delighted to write the foreword for this edited book entitled "Internet of Vehicles and Its Applications in Autonomous Driving". This book highlights the importance that Internet of Things (IoT) technology can offer in taking care of vehicular safety through internetworking and automation. It intends to demonstrate to its readers several useful vehicular applications and architectures that cater to their improved driving requirements. The technology has a wide range of application domains, in which vehicular networking, communication technology, sensor devices, computing materials and devices, IoT communication, intelligent transportation, vehicular and on-road safety, data security and other topics are included.

Internet of Vehicles (IoV) is a trending junction, where IoT meets vehicular technology. The automotive industry continues to evolve and enable the era of intelligent mobility as the autonomous vehicles make impressive strides and new technology emerges before us. The idea of connected and autonomous vehicles continues to be the intersection of automotive intelligence and IoT innovation.

To accomplish all these feats, IoV uses human-centric data and designs. Theory is drawn from IoT and tools are drawn from engineering to provide technologies that improve vehicular and pedestrian safety. As such, a lot of experience, techniques, data, analyses and tools for the advancement of Internet technology are drawn from other fields that include robotics, signal and image processing, virtual and augmented reality, computer science, sensors and actuators, to name but a few fields that contribute to an ever-changing technological crossover.

The range of topics covered in this book is quite extensive, and every topic is discussed by experts in their own field. Overall, the book provides a window to the research and development in the field of vehicular communication in a comprehensive way and enumerates the evolution of contributing tools and techniques. The advances and challenges are discussed with a focus on successes, failures and lessons learned, open issues, unmet challenges and future directions. In this brave new world, it is imperative to provide stakeholders, researchers, practitioners and students with the knowledge of the state of the art and frontiers of intelligent transportation systems, that is vehicular technology.

I highly recommend this book to a variety of audiences, including *academicians*, *commercial engineers* and *researchers* in the fields that use transportation systems, *vehicular communication, Internet of Things, automobile technology practitioners*, communication *technology specialists* as well as *ad hoc communication networks students and scholars*. It is my desire and expectation that this book will provide an effective learning experience, a contemporary update and a practical reference for researchers, professionals and students who are interested in the advances of Internet of Vehicles and its integration into the engineering field.

Federal University of Piauí (UFPI), Teresina, PI, Brazil Joel J. P. C. Rodrigues
Instituto de Telecomunicações, Aveiro, Portugal

Preface

Interconnection of autonomous vehicular networks has led to the evolution of Internet of Vehicles (IoV). Broadly, this can also be termed as a special case of Internet of Things (IoT). Intelligent transportation systems (ITS) is yet another broad set of standards for vehicular communication. The primary objective of all such standards and technologies is to provide improvement in terms of safety and security of moving vehicles mainly on highways. Revolution in the area of mobile communication, computation technology and advancements made in the area of wireless sensor technology have further strengthened the evolution of Internet of Vehicles (IoV).

In IoV architecture, different sensors attached to vehicles collect data related to the traffic conditions on the road and other relevant information. This information is computed as per the application and requirements. All the moving vehicles communicate with each other using vehicle-to-vehicle (V2V) network. As per IoV, three-stage communication will take place, namely V2V, vehicle-to-infrastructure (V2I), and vehicle-to-cloud (V2C). Effective implementation of the above three components using high-speed mobile communication, Internet technologies, and ad hoc networks will increase the application of IoV mainly in smart city, advanced navigation, real-time traffic information, driver and passenger safety, cost-effectiveness, and reduced traffic congestion.

In this book, attempt has been made to present an insight of the importance which IoV can offer in vehicular safety through internetworking and automation. It intends to demonstrate to its readers useful IoV applications and architectures that cater to their improved driving requirements. The application domains have a large range, in which vehicular networking, communication technology, sensor devices, computing materials and devices, IoT communication, intelligent transportation, vehicular and on-road safety, data security, and other relevant topics are included.

The different chapters presented here cover the significant technological advancements that IoV solutions can have in taking care of ITS. The key features of this book are the inclusion and elaboration of recent and emerging developments in various specialization of vehicular communication and their applications in autonomous driving. IoV solutions can directly serve the general public. Their benefits are enormous, and the range of applicability is also significant. Readability level of the book is highly diversified, and it covers the topics being taught through graduate level of engineering colleges to research level throughout the world. The contents of the book will be fully helpful to the students and research scholars. The book showcases a strong focus on practical implementations and simulated runs. The application domain shows a large diversity in which augmented reality, virtual reality, artificial intelligence, communication technology, edge computing, data fusion, privacy challenges, and other topics are discussed. The presented solutions will have the purpose to make everyday life easier throughout the world.

The contents are presented in four main parts: intelligent transport systems and infrastructure, connected vehicles, advance driver-assisted system (ADAS), and evaluation methods and security issues.

The first part talks about overview on ITS in the context of IoV and its necessity in various traffic conditions. The second part focuses on autonomous driving cars, design issues and challenges for IoV, and advances and limitations of medium access control in connected vehicles. The third part highlights ADAS in IoV, augmented reality, and the concept of fusion in ADAS.

Finally, the fourth part presents block chain-enabled security and privacy for IoV, information security issues, and error detection using artificial intelligence.

This book attracted contributions from all over the world, and we would like to thank all the authors for submitting their works. We extend our appreciation to the reviewers for their time and focused review comments. We gratefully acknowledge all the authors and publishers of the books quoted in the references.

Warangal, India — Nishu Gupta
Prayagraj, India — Arun Prakash
Prayagraj, India — Rajeev Tripathi

Acknowledgements

Nishu Gupta I acknowledge the inspiration and blessings of my mother Smt. Rita Rani Gupta, father Prof. K.M. Gupta, and other family members. I am full of gratitude to my wife Smt. Anamika Gupta and son Master Ayaansh Gupta for the patience shown and encouragement given to complete this venture. I wholeheartedly acknowledge the blessings, preaching, and encouragement given by my academic advisor and mentor Prof. Rajeev Tripathi. His real-life lessons and positive attitude towards everything have infused enormous energy within me which has boosted my spirits at different stages. I am highly obliged to my guide Prof. Arun Prakash without whose help, guidance, and support it would not have been possible to bring this book. He has been a major driving force towards this and many other such accomplishments. I extend my heartfelt gratitude to the Principal of Vaagdevi College of Engineering, Warangal, Dr. K. Prakash; Head of the ECE department, Assoc. Prof. M. Shashidhar; and other colleagues and friends for their support and motivation in several ways. I am also thankful to my graduate students Saikrishna B. and Manaswini R. for their assistance at different stages in vivid ways.

Arun Prakash Hardly—if ever—a books is the sole achievement of one person. It is rather a journey where the traveler is dependent on many aids on the road, has to learn the language, has to ask for directions, and often needs a helping hand. It is a great sense of satisfaction with which we are presenting this book. I take this opportunity to thank everyone who made this possible. I express my heartfelt gratitude to Prof. Rajeev Tripathi, Director, MNNIT Allahabad. This work would not have assumed its present shape without his guidance. I am thankful to Dr. Nishu Gupta who put all his effort and has been instrumental in bringing this book. A word of appreciation to the Head of Electronics and Communication Engineering Department and all my colleagues I work with at MNNIT Allahabad. I am also thankful to all my Ph.D. students for their assistance. Finally, I acknowledge my family as the main inspiration to this accomplishment. This list is far from exhaustive; I pray for forgiveness from those I did not mention by name and include them in my heartfelt gratitude.

Rajeev Tripathi I feel pleasure in acknowledging my own student Dr. Nishu Gupta who has worked very hard to bring out this book. He took the initiative, coordinated well with all the authors, and came up with a solid content. I am also thankful to the academia and research fraternity for taking out time to contribute their research efforts in the form of chapter.

Last but not least, *we, the Editors* express our heartfelt gratitude to the publishers and the team behind it for their continued support and cooperation in publishing this book.

Contents

About the Editors

Nishu Gupta is specialized in the field of Vehicular Communication and Networking. He is a Ph.D. in Electronics and Communication Engineering with specialization in designing MAC Protocols for Vehicular Ad hoc Network for enhanced safety in vehicles. Presently, he is a Postdoctoral Research Fellow in the Department of Informatics, University of Oviedo, Spain and also working as Assistant Professor in the Department of Electronics and Communication Engineering at Vaagdevi College of Engineering, Warangal, India. He is an active reviewer of research articles in high impact journals. He has co-authored several books and has more than 38 publications among peer-reviewed journals, book chapters, and conferences proceedings. He has been invited as Speaker, Session Chair, and Conference Chair on various occasions and has been awarded and honored by various academic and research organizations throughout the world owing to his contribution to the fraternity. His research interest includes wireless communication, internet of things, internet of vehicles, autonomous vehicles, edge computing, augmented intelligence, ad hoc networks, vehicular communication, cognitive radio ad hoc network, etc. He is a Senior Member of IEEE and holds active membership of many other professional institutions.

Arun Prakash received his Ph.D. from the Department of Electronics and Communication Engineering, Motilal Nehru National Institute of Technology, Allahabad, India in 2011. He was a Visiting Research Scholar at the University of Waterloo, Canada from September 2008 to February 2009. At present, he is an Associate Professor in the Department of Electronics and Communication Engineering at Motilal Nehru National Institute of Technology, Allahabad, India. His research interests are in the area of wireless and mobile communication, mobile ad hoc networks, vehicular networks (intelligent transportation systems, ITS), cognitive radio networks, and wireless sensor networks. He has published more than 50 papers in international journals and conferences of repute. He is a member of IEEE, IAENG, and IETE.

Rajeev Tripathi received his B.Tech., M.Tech., and Ph.D. degrees in Electronics and Communication Engineering from Allahabad University, India. At present, he is working as Director of Motilal Nehru National Institute of Technology, Allahabad, India where he is also a serving Professor in the Electronics and Communication Engineering department. He worked as a faculty member at the University of the West Indies, St. Augustine, Trinidad, WI, from September 2002 to June 2004. He was a visiting faculty at School of Engineering, Liverpool John Moores University, UK, from May to June 1998 and November to December 1999. He has made pioneering research contributions and has solved a number of open problems. He has published more than 140 papers in international journals and conferences of repute and supervised 18 Ph.D. students till date. He has worked closely with government as well as industry on various problems and has successfully led and completed large projects and programs at national and international levels. His research interests are in the area of wireless and mobile communication systems, mobile ad hoc networks, computer network protocols, sensor networks, and others.

Part I
Intelligent Transport Systems and Infrastructure

An Overview of Intelligent Transportation Systems in the Context of Internet of Vehicles

Matthew N. O. Sadiku, Nishu Gupta, Kirtikumar K. Patel, and Sarhan M. Musa

1 Introduction

Traditionally, the motor vehicles have been affordable means of transportation. As an indispensable part of modern life, vehicles have evolved since their invention in the Second Industrial Revolution. In addition to electric vehicles, smart vehicles have been increasing in number. Internet of vehicles (IoV) is essentially an application of Internet of things (IoT). When all the interconnected devices are vehicles, then IoT becomes Internet of vehicles (IoV), which provides information services, energy-saving emission reduction capability, and driving safety. IoV may be regarded as the evolution of conventional vehicle adhoc network (VANET), which is network that evolved from mobile ad hoc network (MANET). In IoV, vehicles are always connected to the Internet so as to augment the services the users can benefit from when moving in urban areas. Intelligent transportation systems (ITS) integrate advanced communications technologies into vehicles and transportation infrastructure. ITS are basically transportation technologies that support the operations of a state highway system through advanced wireless communication technologies. ITS cover all modes of transportation (road, rail, sea, and air) and provides services that can be used by both passenger and freight transport. It is a system that applies ICT to road transportation including infrastructure, vehicles, and

M. N. O. Sadiku (✉) · S. M. Musa
Roy G. Perry College of Engineering, Prairie View A&M University, Prairie View, TX, USA
e-mail: sadiku@ieee.org; smmusa@pvamu.edu

N. Gupta
Vaagdevi College of Engineering, Warangal, India
e-mail: nishugupta@ieee.org

K. K. Patel
Chemic Engineers, Hitchcock, TX, USA

© Springer Nature Switzerland AG 2021

N. Gupta et al. (eds.), *Internet of Vehicles and its Applications in Autonomous Driving*, Unmanned System Technologies, https://doi.org/10.1007/978-3-030-46335-9_1

users. ITS is an indispensable component of the notion of a smart city transforming cities into digital societies. The smart city concept promotes the growth of social capital through the development of telecommunication, education, and research infrastructure. The range of impacts of ITS may be local, regional, or national [1].

The Japanese carried out the early work on ITS in the 1980s. But Japanese did not coin a specific name for ITS at that time. The concept of intelligent transport systems (or intelligent transportation services) was born in the 1980s when a group of transportation professionals conceived the impact that communication technologies could have on surface transportation. ITS facilitates better public transport services by enhancing safety and easing traffic congestion. ITS is being supported by US Department of Transportation (USDOT) through research, development, adoption, and deployment. The USDOT has provided leadership and cooperation with other federal agencies. The world is at the brink of the most disruptive changes to our transportation system [2].

The ITS Standards Program has teamed up with standards development organizations and participates in international standards harmonization activities. The ITS Standards Program meets the twenty-first century (TEA 21) mandate that the USDOT developed. Nearly 100 standards have been developed so far. It is expedient to use ITS Standards since they are open and non-proprietary. In view of how roads are being used today, transportation agencies and road administrations must prepare for the future. The ITS World Congress is an annual symposium, rotating between USA, Europe, and Asia. It was the first help in Paris in 1994.

2 Enabling Technologies of ITS

Transportation systems play a crucial role in almost all areas of modern life. ITS are basically transportation technologies that support the operations of a state highway system through advanced wireless communication technologies. They are being used to improve the safety, comfort, and productivity of our surface transportation systems. They can potentially revolutionize mobility, changing everything from the way we move to how we design vehicles. ITS are innovative solutions that address modern transportation problems. A typical ITS is shown in Fig. 1 [3].

ITS uses sensing, analysis, control, and communications technologies as well as management strategies for ground transportation in order to improve safety, mobility, and efficiency. There is a range of technologies that enable the development or application of ITS. These enabling technologies can be divided into several classes which include [4, 5]:

1. *Communication:* Various forms of communication include
   (a) Wired communication: Fiber optic, twisted pair wires
   (b) Wireless communication: UHF, VHF, WiMAX, GSM, Infrared, microwave, radio, cellular technology

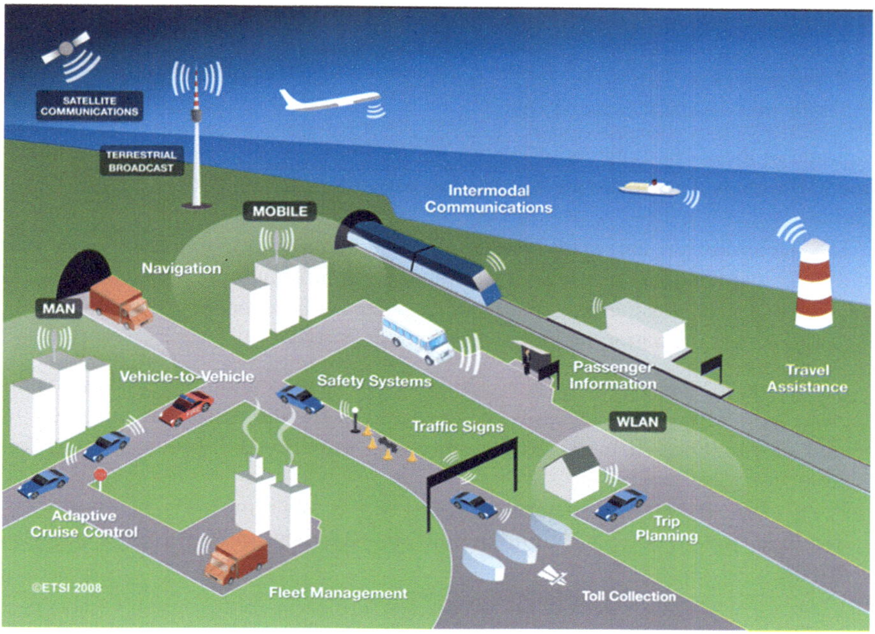

Fig. 1 A typical ITS [3]

2. *Sensing technology:* A variety of sensors have been developed. Weather sensors such as rain gauges have been put to practical use.
3. *Software*: This is necessary to implement and optimize the desired behaviors in these systems.
4. *Data acquisition:* Data collection devices gather information from the transportation systems. Cameras and RFID scanners are used for data collection. Traffic can be monitored using inductive loop detectors and traffic sensors such as radar and video image detector (VID)
5. *Data processing*

ITS often operates in an invisible manner, with buried cabling or wireless communications. Computers, electronics, satellites, sensors, GPS, and Internet are playing an increasingly role in the transport systems.

3 Components of ITS

ITS typically consists of the following components [6]:

- *Active traffic management* (ATM)
- *Traffic cameras*(CCTV)

- *Highway advisory radios* (HAR)
- *Road/weather information systems*(RWIS)
- *Ramp meters*
- *Traffic management centers*(TMC)

Some of these components of ITS are shown in Fig. 2 [7]. The figure also shows how ITS uses ICT to create an information network based on people, vehicles, and roads.

ITS technologies are many including vehicle magnetic signature detection, automatic incident detection, parking guidance and information systems, automatic road enforcement, weather information, etc. Some of these technologies are already deployed across the nation include the following [8–10]:

- *Electronic toll collection* (ETC)**:** This enables identification of registered vehicles. It directly debits accounts of registered users and alerts law enforcers in case of violations. It increases the efficiency, mobility, and provides faster toll collection.
- *Ramp meter* (RM)**:** This meter is installed on freeway ramp.
- *Red light camera* (RLC)**:** Helpful in traffic management and controlling unlawful driving.

Fig. 2 Components of ITS [7]

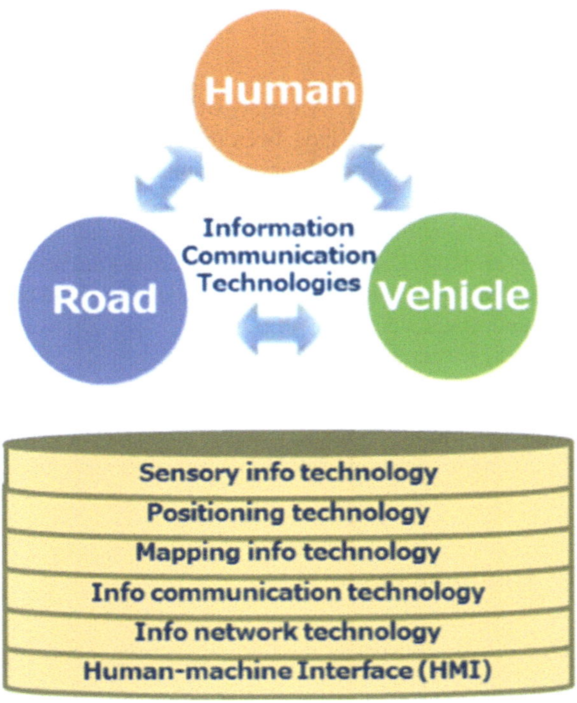

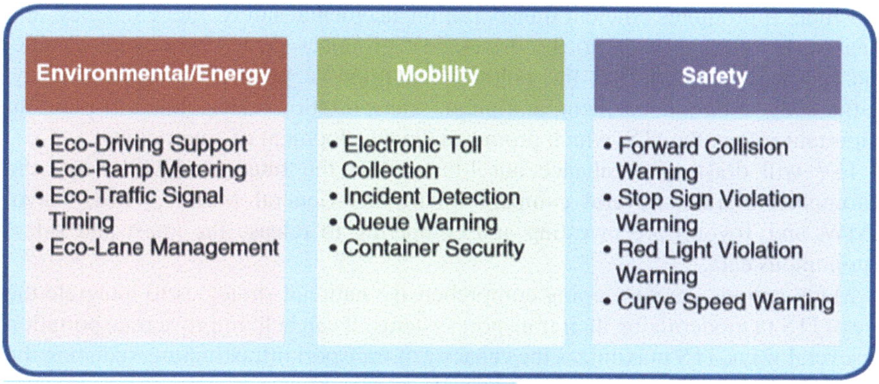

Fig. 3 Examples of ITS applications [11]

- *Transit signal priority* (TSP): TSP extends the duration of green signals for public transportation vehicles if necessary.
- *Traveler information systems* (TIS).
- *Global positioning system* (GPS).
- *Automatic road enforcement.*

Some of these examples of ITS applications are illustrated in Fig. 3 [11].

Many states and cities are increasingly adopting ITS as smart transportation systems. The government (local or state) are spending millions of dollars to deploy ITS. Here, we follow the example of implementing ITS in California. At the state level, Caltrans design, construct, maintain, and operate the state highway system and the intercity passenger system. The first priority should be to improve the existing system and protect the huge investment that is already made. A decision needs to be made regarding which ITS services to implement. Elements of the ITS can be designed, installed, and integrated. Full project deployment will take time, money, and visionary leadership [12]. Government may choose to privatize ITS projects and the cost of deployment will be recouped by the private company from tolls paid by the road users.

4 Benefits

As an application of the Internet of things, Internet of vehicles has the technical characteristics of the Internet of things, an emerging technology which envisions to make human lives smarter. The prime goal of IoV is inter-connectivity among smart vehicles. IoV helps a vehicle to connect to another vehicle using a wireless network.

The IoV is designed to guarantee road safety, improve traffic efficiency, and provide entertainment service due to its capability to share vehicular messages in real-time. It promotes efficient transportation, accidental safety, fuel efficiency, and pleasant travel experience to the driver. IoV promises great commercial benefits and research value. It has the potential to improve people's travel efficiency, traffic congestion management, and public transportation management. It plays an important role in the ITS, which promotes the development of smart cities.

IoV will drastically enhance our lifestyles in the future and will boost the automobile market. Several companies (such as General Motors, Ford, Volvo, BMW, and Toyota) are investing and competing to release the latest and safest autonomous cars.

Many nations are developing comprehensive national strategies to integrate the use of ITS in modernizing their transport systems. It can help improve transportation in several ways. ITS maximizes the capacity of transport infrastructure, reducing the need to build additional highway capacity. Government leaders favor ITS because of its promise to significantly improve the efficiency of existing transportation systems. Other benefits of ITS over traditional transportation system include the following [13].

- *Improving traffic safety*: Safety is one of the major driving forces behind the implementation of ITS systems. The applications of the ITS are helping with confronting today's traffic challenges such as unsafe speeds, accidents, dangerous weather conditions, and traffic congestion. ITS enhances safety, reduces risk, increases comfort, improves mobility, optimizes traffic efficiency, promotes sustainability, and minimizes traffic problems in a number of situations.
- *Reducing infrastructure damage*: Overload vehicles can put great strain on the road and damage it. Weigh-in-motion systems measure weight of vehicles as they move and identify overloaded vehicles. These systems make enforcement easier.
- *Traffic control*: ITS allows traffic lights to respond to changing patterns automatically rather than on a fixed schedule, prioritize specific forms of traffic, and keep traffic moving smoothly.
- *Parking management*: Smart parking management can put a check on unauthorized parking and overstaying drivers.
- *Gathering traffic data*: With increased connectivity among vehicles, systems, people, mobile devices, and other entities, unprecedented amounts of data are being generated. Homeland security can use the data available from ITS.
- *Environmental impact*: Drivers and businesses want transportation system that is safe, cost-effective, reliable, and respectful of the environment. ITS plays a major role in the reducing the negative environmental impact of transportation and achieving sustainable development requirements. It can also contribute to poverty reduction by improving the travel costs. It makes transport more user friendly. It holds the promise of sustainability and enhances productivity. The possibilities are endless.

5 Challenges

Despite all the benefits, many nations cannot invest much in ITS. This is partly due to the challenges involved in deploying ITS. The challenges include the following [14].

- *Personal privacy*: The use of ITS brings concerns that user's privacy may be negatively affected. Privacy is a primary concern when tracking people, vehicles, and goods. Data privacy and protection can be sensitive issues in a wireless communications environment where personal information such as locations traveled and driving behavior can be of interest to law enforcement agencies [15]. The data may be filtered for privacy or security. Computer hackers may target ITS and cause collisions and traffic disruption.
- *Standardization*: The proliferation of ITS projects has inevitably led to problems of compatibility and interoperability within and across nations. Developing new ITS standards at national and international levels is important [1].
- *Interoperability*: This deals with how to communicate effectively with other parts regardless of where they are built. As our environments become more connected, interoperability will be more critical than ever before.
- *Awareness*: Most people including governments, individuals, ITS practitioners (such as urban designers, town planners, landscape architects, and surveyors) do not realize that they need to be able to argue convincingly for investment in these technologies. By engaging the right audiences, we can facilitate the transition from ITS adoption to large-scale deployment. Communication, education, and training are important activities that will facilitate awareness.
- *Politics*: Politicians are failing to enact the policies and incentives that are needed to adapt the ITS and create better, smart cities. ITS community will fail to deliver transport that serves our society better if the self-serving politicians do not consider the wider interests of society. Politicians and governments should take a long-term view in providing funding.
- *Complexity*: Due to the complexity of IoV, its adoption in urban environments will take longer. Factors such as inefficiency in determining the exact vehicle position hinder the market growth to a certain extent.

ITS also faces jurisdictional challenges, such as which level of government (federal, state, etc.) has responsibility for ITS deployments. These challenges are not simple issues. None of them are insurmountable, and some nations have overcome them.

6 Recent Developments

Recent advances in technology have encouraged further development of ITS. Wireless sensor networks (WSNs), big data, cloud computing, SDN, VANETS, and visualization have been merged into ITS.

- *Wireless sensor networks*: WSN is a promising technology providing ITS to address the traffic problems of congestion and safety. The major advantage of WSN is its low cost, portability, and distributed intelligence [16]. For example, wireless sensors can be deployed in parking garages, in the roads, or mounted above the lane of traffic they are monitoring. Smart vehicles are equipped with sensors which have the capability to sense, process, and distribute data from the environment.
- *Big Data*: Effective use of the abundant data generated by ITS will benefit the driver and improve road safety. Analyzing the big data produced can be used to make ITS safer, more efficient, and profitable. Big data analytics in ITS include personal travel route plan, and rail transportation management [17]. Big data related technologies support data-driven modeling and provide new ideas and tools for smart management and decision-making.
- *Visualization*: This refers to the ability to display assets graphically. It is one of the most important factors for ITS asset management and decision-making. The visualization capability could include map viewing, ITS visualization, and wireless network visualization [18].
- *Cloud computing*: This is the delivery of computing as a service over a network such as the Internet. Cloud computing can be applied to ITS to improve transport outcomes such as road safety, transport productivity, travel reliability, informed travel choices, and environment protection [19].

7 Conclusion

The ever-increasing need for mobility has overflooded cities with vehicles, causing increasing traffic congestion, delays, accidents, vehicle emissions, and making.

Transportation system reaches the limits of its existing capacity. Transportation is changing rapidly with the advances in technology to solve traffic problems. ITS applies ICT technologies to the transport sector and integrates people, roads, and vehicles and significantly contributes to improve road safety, efficiency, and comfort, as well as environmental conservation. Although ITS is often associated with road traffic, other transport modes (such as air traffic) are becoming ITS. ITS is widely accepted and used worldwide. In developed countries, ITS is already installed on a large number of highways. In developing countries, the installation of ITS is increasing. It is increasingly being used to improve traffic in rapidly growing cities. It appears that the introducing ITS will bring huge benefits for cities. New smart transportation models and technologies are emerging globally. The days of adding more lanes to ease congestion are gone. New technologies such as ITS are designed to improve surface transportation.

References

1. Smith, B. L., & Venkatanarayana, R. (2005). Realizing the promise of intelligent transportation systems (ITS) data archives. *Journal of Intelligent Transportation Systems, 9*(4), 175–185.
2. History of intelligent transportation systems (ITS): Connected vehicles and smart cities. Retrieved from https://www.its.dot.gov/presentations/2017/AVCV_ITSHistory.pdf
3. The future of Intelligent Transport Systems (ITS). (2011). Retrieved from https://mubbisherahmed.wordpress.com/2011/11/29/the-future-of-intelligent-transport-systems-its/
4. Intelligent transport systems: EU-funded research for efficient, clean and safe road transport. Retrieved from http://www.greendigitalcharter.eu/wp-content/uploads/2012/11/2010-European-Commission-Report-on-Intelligent-Transport-Systems.pdf
5. What is intelligent transportation system? Its working and advantages. Retrieved from https://theconstructor.org/transportation/intelligent-transportation-system/1120/
6. Intelligent transportation systems (ITS) operations. Unknown Source.
7. Intelligent transport systems (ITS): Introduction guide. Retrieved from http://www.jsce-int.org/system/files/ITS_Introduction_Guide_2.pdf
8. Pina, M. ITS research fact sheets—Benefits of intelligent transportation systems. Retrieved from https://www.its.dot.gov/factsheets/benefits_factsheet.htm
9. Intelligent transportation system. Wikipedia, the free encyclopedia Retrieved from https://en.wikipedia.org/wiki/File:GDS-display.jpg
10. Intelligent transport systems: Reference material for competence. Retrieved from https://www.eltis.org/sites/default/files/ITS_Telematics_6.pdf.
11. Chowdhury, M., & Dey, K. (2016). Intelligent transportation systems—A frontier for breaking boundaries of traditional academic engineering disciplines. *IEEE Intelligent Transportation Systems Magazine, 8*(1), 4–8.
12. Golob, J. M., Stecher, C. C., & Felkin, C. (2003). California statewide intelligent transportation systems plan evaluation case study of conformity with national intelligent transportation systems architecture. *Transportation Research Records: Journal of Transportation Research Board, 1826*(1), 1–6.
13. What are the benefits of intelligent transportation systems? Retrieved from https://advanceaccess.ie/benefits-intelligent-transportation-systems/
14. ITS 2015–2019 strategic plan. (2003). US Department of Transportation.
15. Lederman, J., Taylor, B. D., & Garrett, M. A private matter: The implications of privacy regulations for intelligent transportation systems. *Transportation Planning and Technology, 39*(2), 115–135.
16. Peng, Y., Li, J., Park, S., Zhu, K., Hassan, M. M., & Alsanad, A. (2019). Energy-efficient cooperative transmission for intelligent transportation systems. *Future Generation Computer Systems, 94*, 634–640.
17. Zhu, L. (2019). Big data analytics in intelligent transportation systems: A survey. *IEEE Transactions on Intelligent Transportation Systems, 20*(1), 383–398.
18. Fries, R., Anjuman, T., & Chowdhury, M. (2013). Selecting an asset management system for intelligent transportation systems. *Public Works Management & Policy, 18*(4), 322–337.
19. Bitam, S.,&Mellouk, A. (2012). ITS-Cloud: Cloud computing for intelligent transportation system. *Communications Software, Services and Multimedia Symposium 2012 (Globecom)*, 2054–2059.
20. Brief introduction to intelligent transportation system, ITS. https://www.freeway.gov.tw/UserFiles/File/Traffic/A1%20Brief%20introduction%20to%20Intelligent%20Transportation%20System%20ITS.pdf

Intelligent Transportation Systems and Its Necessity in Various Traffic Conditions in Indian Scenarios

Sankar. P and Gayathri Voorandoori

1 Introduction

Vehicular ad hoc network (VANET) is a wireless communication network system, which is part of the mobile ad hoc network. The routing of data packets is a critical factor in the VANET communication system, since the mobility of the nodes is high. It uses various routing protocols to ensure proper delivery of the data packets to the destination. In this chapter five real time scenarios are simulated that operate under the VANET propagation model. For the simulation of those scenarios, two protocols are used, one from the reactive type, called as ad hoc on-demand distance vector, and the other from the proactive type, called as optimized link state routing. These simulations and analysis can help in providing solutions to various problems in traffic and road traffic management. The main objective of this chapter is to provide solutions to problems like traffic congestion at the junctions, accident, and blocking of the emergency vehicles that prevail in the current traffic control system and also in road transport in India. This chapter also aims at the comparison of two routing protocols and suggests a suitable protocol that can be used for those scenarios.

This chapter is organized into seven sections; each one deals with specific aspects. The related works are described in Sect. 2. Section 3 briefs the VANET propagation system. The Sect. 4 enumerates the concept of how the routing of data happens in the OLSR and AODV protocols. Section 5 explains the five scenarios considered in the chapter. The results obtained are evaluated in Sect. 6. The conclusive statements and future scope are stated in Sect. 7.

Sankar. P (✉) · G. Voorandoori
SV Engineering College, Tirupati, India
e-mail: sankar.padmanaban@svcolleges.edu.in; gayathri.v1@svcolleges.edu.in

© Springer Nature Switzerland AG 2021

N. Gupta et al. (eds.), *Internet of Vehicles and its Applications in Autonomous Driving*, Unmanned System Technologies, https://doi.org/10.1007/978-3-030-46335-9_2

2 Related Works

In the recent past, various proposals have been made for efficient traffic and road transport management. Some of the proposals and the constraints in the proposals are explained in this section. Qi Wang et al. [1] stated that intelligent transportation systems (ITS) are based on the acquisition of urban traffic information in time and accurately. This is performed through the wireless traffic information service system. In this system, vehicles are equipped with a wireless sensor module, which has the provision to communicate. Roadside nodes know the traffic condition by analyzing the data collected from the vehicles. The internet can also be used to monitor the traffic status. This concept was proposed by Hu Lingling et al. [2]. According to this concept, a vehicle is allotted with a unique identification number and it is written electronically in the RFID tag. When the vehicle reaches the base station area, the RFID tag is read, and the internet facility is provided. Using GPS, the traffic status is obtained through the internet. Thereby, drivers can take an alternative free path to reach their destination easily. Pravin P Ashtankar et al. [3] proposed a solution to prevent accidents. The central idea of this system is to enable vehicles within each other's proximity to be aware of their own location and then estimate their position with respect to other vehicles. This is accomplished with the help of two main technologies called global positioning system (GPS) and radio detection and ranging (Radar). Navin Kumar et al. [4] stated that traffic information can be broadcasted using visible light communication. According to their proposal, the currently available traffic light controller (TLC) made with the light emitting diode (LED) can be used to broadcast the traffic data. The vehicle in front of the TLC receives the data with the help of photodiode. This is a line-of-sight (LOS) communication. The vehicle which receives the information can relay it to the vehicle behind it, using the brake lights on the rear panel. This leads to an ad hoc V2V communication. LOS communication between the V2I and I2V is also possible, to learn about the present traffic condition. This system is proved to be effective if the communication range is less than 40 m. As LEDs and the currently available TLCs have been used, the system is cost effective. Another major proposal that paves the way for an excellent traffic management was given by Mounib Khanafer et al. [5]. The wireless sensor network for ITS collects and communicates information to organize the traffic intersections. This architecture depends on a fixed infrastructure composed of intersection units and roadside units. Vehicle-to-vehicle and vehicle-to-roadside units' communications are supported. Vehicle units (vehicles equipped with sensors) send vehicle parameters (speed, direction, and location) to the roadside units. The roadside units aggregate the received data and then transfer it to the intersection units. Finally, the aggregated data reaches the strategy sub-system. Nazmus S. Nafi and Jamil Y. Khan [6] suggested a solution to avoid waiting for a long time at the junction for the green signal. They proposed that the vehicles with the on-board unit (OBU) will communicate with the roadside unit (RSU). The RSU will get the status information of the TLC and send it to the OBU in range. By knowing this information, the driver can decide whether to accelerate or decelerate to avoid wasting time and fuel at the red signal.

The congestion at the intersection can be reduced by dynamically changing the TLC phase. This concept was proposed by Maythem K. Abbas et al. [7]. They stated that using five devices, congestion can be reduced, to provide safety and comfort in transportation. A coordinator device (CoD) controls the communicating devices like the TLC and the roadside equipment (RSE). The RSE communicates with vehicles as well as with the TLC. A vehicle VANET device is fitted in the vehicle for receiving the data. The VANET road belt is a strip of sensors placed on the road to find the traffic level. After finding the traffic level, the data is transmitted to the RSE and using this data, the TLC phase is changed dynamically to reduce the traffic congestion. Yuanguo Bi et al. [8] explain about the protocol used for the exchange of information between the vehicles. They addressed the disadvantages of various protocols that are used in vehicular ad hoc networks and proposed a position based multi-hop broadcast protocol for emergency message dissemination in inter-vehicle communication. This protocol selects the candidate vehicle for forwarding an emergency message, which is selected according to its distance from the source vehicle in the message propagation direction. Koushik Mandal et al. [9] presented an intelligent traffic congestion monitoring and measurement system to monitor and measure the road traffic congestion, using the probe vehicle. Real time traffic data can be analyzed with the probe vehicle concept. The system provides a platform to analyze the traffic condition and congestion pattern. RFID and GSM technologies are used in this system. The system will use roadside active wireless devices to collect signals from the active RFID tags attached to the probe vehicle. The goal is to implement a system that would trace the travel time of the probe vehicle as it passes the roadside devices and calculate an average trip time. Feng Li et al. [10] proposed an intelligent vehicle routing system on real time traffic flow information of the road network and real time in-transit goods' status information. The RFID reader of the system reads active RFID tags attached to goods in-transit and extracts the data. The traffic flow information of observed roads and the vehicle location information are fed back by the drivers. Historical data are used to find the traffic condition of unobserved roads. RFID technology can also be used as an extra sensing module that is used in complex surroundings such as dense underground tunnels and urban areas [11]. The benefits of RFID technology can be seen in existing areas where the technology is applied. One of the benefits of an RFID transponder is that no line of sight is required and, more importantly, no human intervention is required [12].

3 Vehicular Ad Hoc Network

The VANET enables communication between vehicles and between the vehicle and the nodes on the roadside, using the allotted frequency without the use of an access point. It shares some common characteristics with the MANET. Both are characterized by the movement and self-organization of the nodes. There are various protocols existing for MANETs, but further studies are required to evaluate the suitability of the existing protocols for VANET. It uses technologies like ZigBee, WiMax, IEEE 802.11, etc., for communication between vehicles on active mobility. The main target in VANET is to build the safest and most useful roads. The factors that are responsible for the adoption of VANET are safety applications that require low latency, secure, and the faster growth of multimedia applications. Recently, much research has been undertaken in this field because many believe that the speed and dynamics of its vehicular network make the VANET more unique and specific than the general MANET. But actually, it is developed as a part of ITS and leading by the reference communications stack ISO/ETSI [13]. Two types of VANET communications are there in VANET each have their private constraints within various scenarios. In addition, each vehicle is again intermittently transmitting the message to each other that used to exchange their current states and surrounding information. Due to this situation, it consumes a high bandwidth from limited VANETs spectrum [14]. The chapter gives an overview on the existing standards in vehicular networking and highlights the upcoming trends in an integrated infrastructure based on the internetworking of different technologies [15]. In another work the researchers used the floating car data information for providing solutions efficiently by using the dedicated short-range communication [16].

4 Routing Protocols

The two routing protocols OLSR and AODV have been explained in brief in the preceding subsections.

4.1 Optimized Link State Routing

The OLSR is based on link state routing protocols. It is developed for mobile ad hoc networks. It operates as a table-driven protocol, i.e., each node maintains a table with the routes to all the destinations in a network. Each node selects a set of its neighbor nodes as MPRs (multi-point relays). Control messages are used to update the link state information between the nodes. The packets to the destination are forwarded by the MPRs. The routing table is formed with the help of the information obtained through the control messages.

4.2 Ad Hoc On-Demand Distance Vector Routing

The AODV is a reactive protocol suitable for wireless ad hoc networks. It is based on distance vector routing. Each node finds the route to the destination only when it is required. It is loop free and avoids the counting to infinity problem. It uses the destination sequence numbers to find the recent route to the destination. It uses three types of messages. They are route request, route reply, and route error. The source node sends the route request to the destination through its neighbors, who forward the request to the destination. The route reply will be sent to the source by the destination through the minimum hop count path. Error messages are generated if any node moves away from the network, or if there is any failure in the link.

5 Scenarios

Two real time video clippings obtained from the traffic police video collections are shown in Fig. 1a, b. It clearly states the accidents and irregularities in the Indian traffic system. Some of the reasons for congestion, collision, etc. are due to the lack of awareness, overspeeding, traffic signal violation, and jaywalking.

The scenarios used for simulation of the traffic conditions using the two protocols are as followed in the subsections.

5.1 Congestion

The traffic density in cities increases day by day. It is becoming very hard to control the traffic that leads to traffic congestion. If there is a provision to take a free left, without knowing the traffic level at the junction, drivers choose the straight path

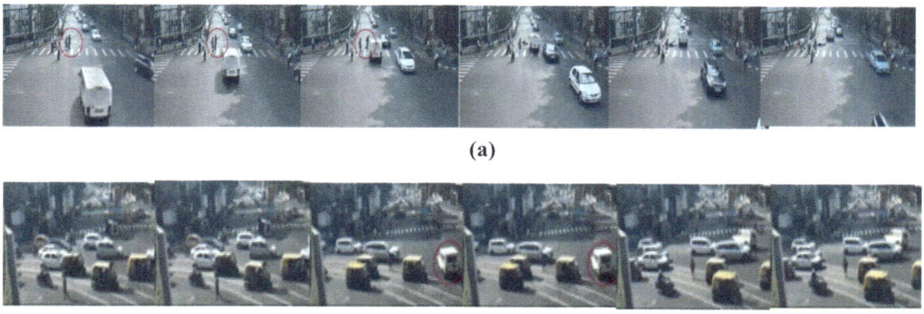

(a)

(b)

Fig. 1 (**a**) Jaywalking. (**b**) Traffic rule violation leading to congestion

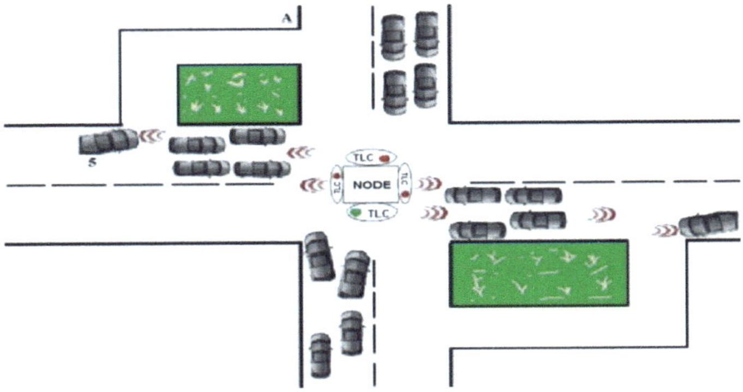

Fig. 2 Congestion scenario

and get stuck at the junction. In this chapter, a solution to reduce congestion and unnecessarily waiting for the traffic signal while having an alternative path to their left is proposed. The considered scenario is shown in Fig. 2. Vehicle no. 5 wants to go to place A. The place is 1 km if we use the straight path and it is 1.5 km if we use the left path. If we use the straight path, we have to wait for approximately 2 min for the signal to turn green, and the allowed speed is 25 km/h. If we choose the left path there is no need to wait for the signal and no speed limit.

Assume the speed and the distance covered as in the following calculation:

1. Time taken to reach place A by the vehicle through the straight path:

   Let T_1 be the worst case of waiting time at the junction for signal:

$$T_1 = 2.5 \text{ min} \tag{1}$$

   Let T_2 be the driving time to place A:

$$T_2 = 1 \text{ km}/25 \text{ km/h} \tag{2}$$

$$T_2 = 2.4 \text{ min} \tag{3}$$

   Let T be the total time taken:

$$T = T_1 + T_2 = 2.5 + 2.4 = 4.9 \text{ min} \tag{4}$$

2. Time taken to reach place A by the vehicle through the left path:

   Let T_3 be the driving time to place A:

$$T_3 = 1.5 \text{ km}/25 \text{ km/h} \tag{5}$$

$$T_3 = 3.6 \text{ min} \tag{6}$$

This proposed system makes us reach the destination fast and reduce the traffic congestion. It can be implemented by sending information about the traffic level at the junction to vehicle no. 5 and by knowing the information; the driver will take the left path that is available and reach the destination fast. The data sent is in the video form. This is like live video streaming without using the satellite. The node at the junction gets the video from the cameras for all the four directions separately and encodes the video and transmits it to the vehicles. Vehicles in a particular direction will get that direction's videos. Even though a single node transfers the video to all the directions, there will not be any interference as done by space division multiple access (SDMA) concept. According to the given scenario, consider the full figure as a cell, and the roads in four directions as four sectors in the cell. Each sector will be served by the different beams from the smart antennas at the nodes.

5.2 *Emergency*

Heavily congested traffic conditions at the junction not only block the vehicles to take a free left, but also create problems for emergency vehicles. Even though emergency vehicles have the siren facility, it increases sound pollution and adds to the pressure of the drivers of other vehicles. So, a system has been proposed that provides the way for emergency vehicles to move in an effective manner. Here, it is assumed that all the vehicles are connected through the VANET in our system. Emergency vehicles send prior information to other vehicles which get stuck at the junction and also to the traffic light controller (TLC) as shown in Fig. 3. Upon receiving the signal from the emergency vehicle, the TLC will turn on the green light in the direction in which the emergency vehicle is approaching, irrespective of the direction in which the green light was on currently.

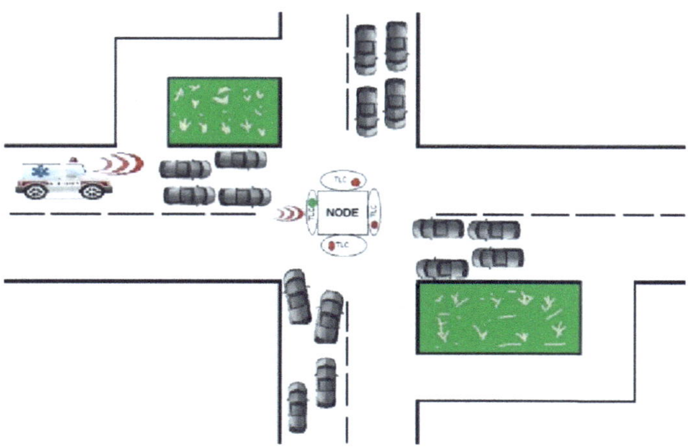

Fig. 3 Emergency scenario

This communication between the vehicles eliminates the necessity of the use of the siren and provides faster access of roads to the emergency vehicles.

5.3 Collision Avoidance

For the past few years, accidents have been increasing day by day. The reasons for the occurrence of such accidents are many. A few reasons are drunken driving, overspeeding, etc., in this fast world, it has become necessary to have an efficient transportation system, which prevents the occurrence of such accidents. The proposed system to prevent accidents using the VANET can be applied at the junction of three roads, where there is no provision of traffic lights as shown in Fig. 4. If the vehicle wants to turn left or right, it will generate the alert signals to the vehicles near the junction, asking them to wait till it crosses the junction and take the corresponding direction. The alert signals are transmitted by comparing the junction crossing time of the vehicles.

Let two vehicles be $V1$ and $V2$.

Let the speed of $V1$ and $V2$ be $S1$ and $S2$, respectively.

Let $X1$ and $X2$ be the distance between the current position of vehicles $V1$ and $V2$ and the junction.

The time taken by $V1$ to cross the junction is $T1$

$$T1 = X1/S1 \tag{7}$$

The time taken by $V2$ to cross the junction is $T2$

$$T2 = X2/S2 \tag{8}$$

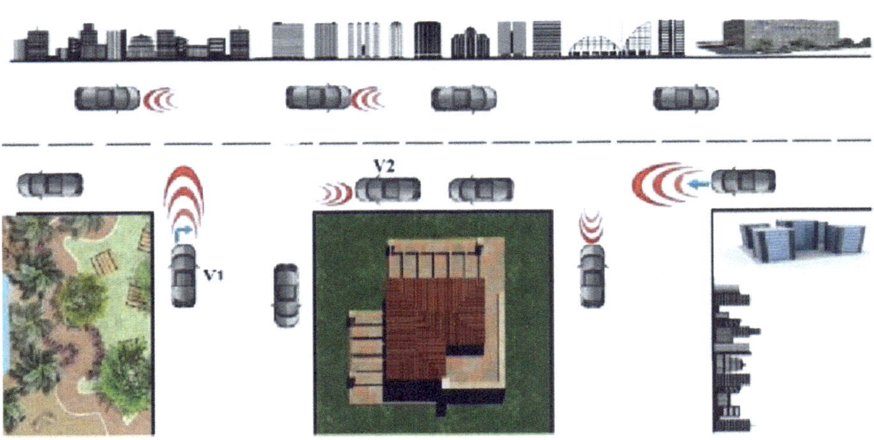

Fig. 4 Collision avoidance scenario

The time taken to cross the junction by the vehicles will be shared between the vehicles.

If ($T1 > T2$) or ($T2 > T1$), then there will be less possibility of collision and there is no need of an alert signal generation.

If ($T1 = T2$), the alert signal is generated and transmitted to the vehicles nearing the junction.

Upon receiving the alert signal the vehicles will wait. If they want to move urgently, they will send the signal in return to the alert signal generated by the previous vehicle. On receiving the return signal, the vehicle should wait for the other vehicle to cross the junction.

5.4 *Collision Detection*

Although there are various ways to avoid accidents, overspeeding of vehicles, drunken driving, and driving without concentration may cause accidents. The information about the occurrence of such accidents reaches the emergency and rescue department only through phone calls from the injured person or from a passerby. This may involve an unnecessary waste of time during an emergency. A system has been proposed to avoid these phone calls. The VANET based system contains the nodes placed along the roadside at specific distances. The tri-axial accelerometer sensors placed in the vehicles will sense the accidents and will transmit the information about the accident to the nearby reachable nodes on the road. After receiving the information, the roadside node broadcasts the information as shown in Fig. 5.

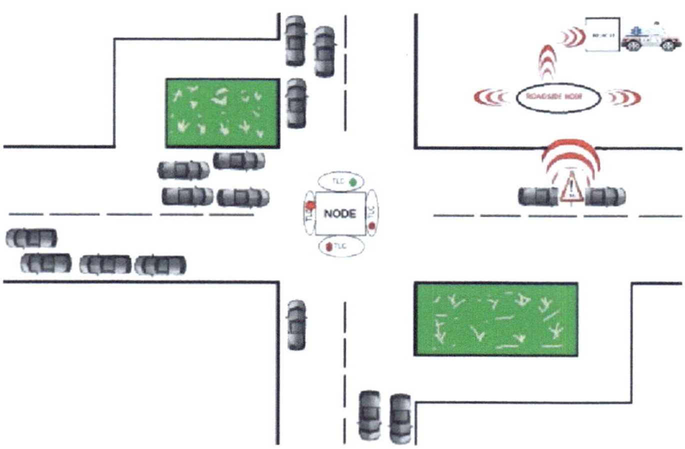

Fig. 5 Collision detection scenario

Fig. 6 Overtake scenario

On receiving the data from the roadside node, the nearby rescue department will send the rescue vehicle to the place from where the information was received.

5.5 Overtake

In addition to all the above systems, another system has been proposed that can be adopted to ensure the safe overtake by a vehicle on the highway. In a nonlane road system on the highway, a vehicle can overtake another one either on the left or right. This is generally called as the random waypoint movement. Currently, drivers signal the vehicles at the back using indicators or gestures by hand. The proposed system involves the communication between the vehicles through the VANET, as shown in Fig. 6.

If a vehicle wants to overtake another one on the highway, it sends the information to the vehicle about the direction in which it intends to overtake. On receiving the information, the vehicle waits for the other vehicle to overtake it in the specified direction. This avoids an accident that may occur during the overtake and ensures a safe journey.

6 Simulation Result

In routing of VANET, the protocols have been classified into five categories. They are multicast based, geocast based, cluster based, position based, and topology based. This taxonomy of VANET protocols is shown in Fig. 7. In this chapter the

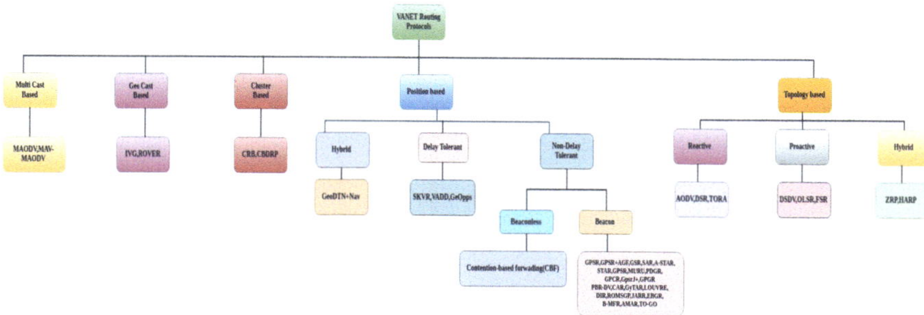

Fig. 7 Taxonomy of VANET routing protocols

authors considered the topology-based protocols such as OLSR and AODV for the simulation of various scenarios, as the topology keeps on changing due to the traffic conditions in India.

The simulation of various scenarios is carried out using C++ based discrete event simulator, NS2. Figures 8a–e and 9a–e show the graphs obtained for packet delivery ratio (PDR) and average delay, respectively, for all the scenarios. For the congestion scenario in Fig. 9a the simulation shows that the OLSR protocol is prominent compared to AODV, since there is a huge packet loss in the transmission using the AODV protocol. It is observed that the PDR decreases gradually from 0.435 GHz to 2.45 GHz and at 3 GHz, it suddenly drops to a lower value of less than 60. The average delay was found to increase with the frequency and due to the large packet loss at 3 GHz, it drops to 1 ms. At 0.435 GHz, it can be seen that the PDR is the maximum, and the average delay is the minimum with a smaller number of packet losses. Since the delay with the minimum packet loss is the critical factor to be avoided in video streaming applications, the scenario operating under 0.435 GHz can be adopted in real time. In order to avoid jitter at this frequency the average delay value must be maintained below 3.55 ms. In the emergency scenario, the PDR for both OLSR and AODV is found to be exactly 100% for low frequency values, and it decreases with higher frequency. The average delay values are found to be optimum at 0.435 GHz and 0.915 GHz with the AODV protocol; when compared with the OLSR protocol and at 3 GHz, the average delay is high due to the large end-to-end delay value. In all the scenarios, if applied in real time, 0.435 GHz or 0.915 GHz frequency can be used with the AODV protocol, because the PDR is 100% and the average delay is the minimum. For the collision avoidance scenario, it is observed from the obtained result that the PDR and the average delay of the AODV protocol oscillate with frequency. The PDR of the OLSR protocol increases with frequency and at 3 GHz, it attains the maximum value and the average delay value is the minimum. Hence, in real time this scenario can be implemented at 3GHz with the OLSR protocol. For collision detection, the PDR was found to be 100% for all frequencies except 3GHz, and the average delay is the optimum with the AODV protocol. When simulated with the OLSR protocol, both metrics do not have fair

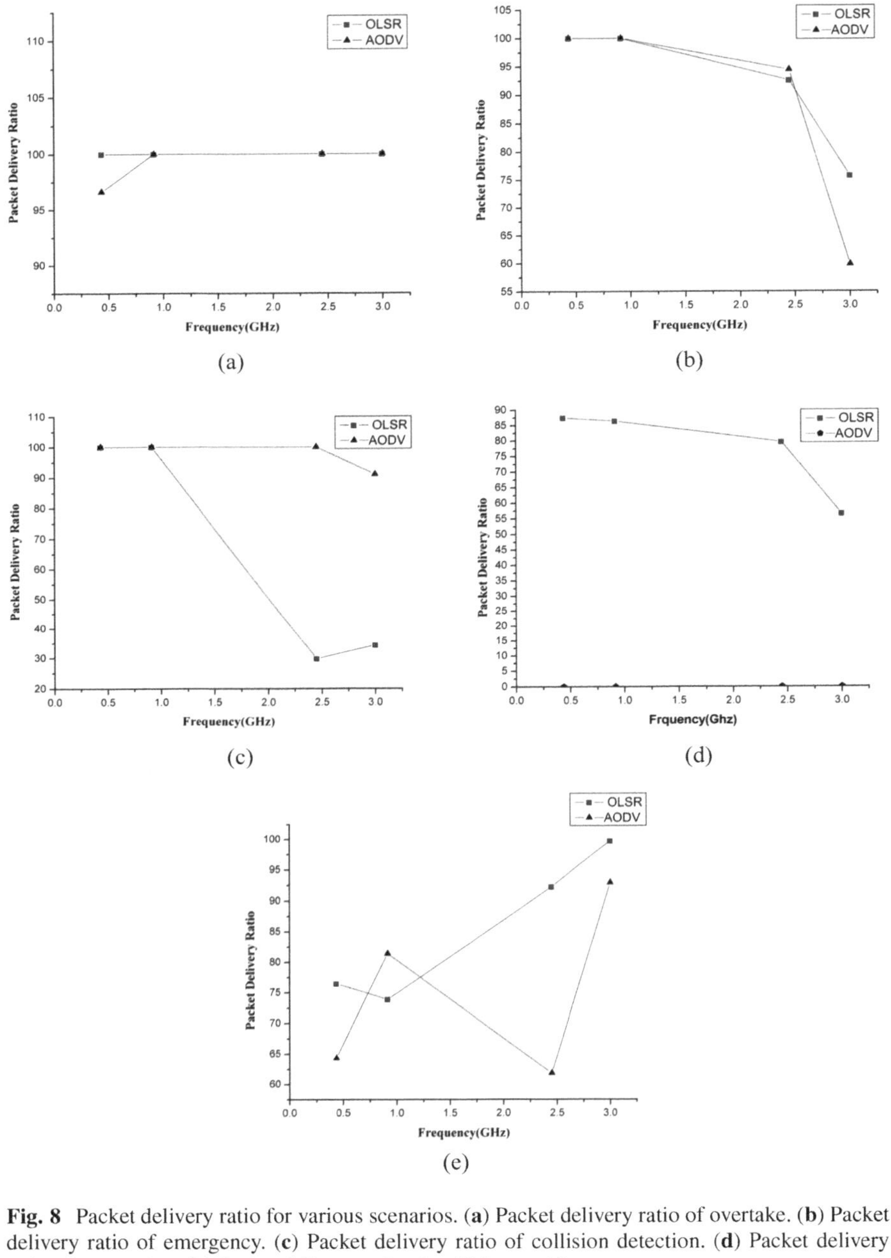

Fig. 8 Packet delivery ratio for various scenarios. (**a**) Packet delivery ratio of overtake. (**b**) Packet delivery ratio of emergency. (**c**) Packet delivery ratio of collision detection. (**d**) Packet delivery ratio of congestion. (**e**) Packet delivery ratio of collision avoidance

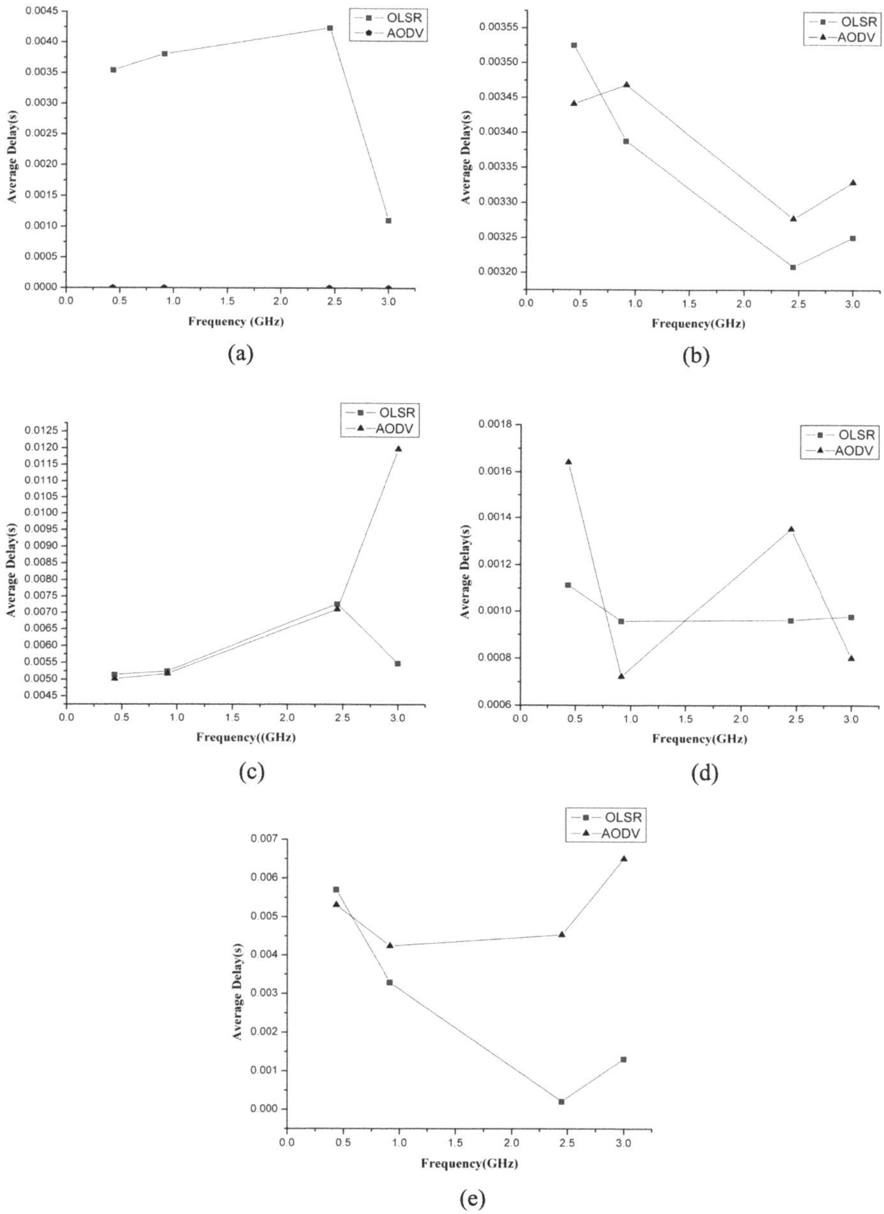

Fig. 9 Average delay of all scenarios. (**a**) Average delay congestion. (**b**) Average delay overtake. (**c**) Average delay emergency. (**d**) Average delay collision avoidance. (**e**) Average delay collision detection

Table 1 Suitable protocol and frequency

Scenarios	Protocol	Frequency (GHz)
Congestion	OLSR	0.435
Emergency	AODV	0.435/0.915
Collision avoidance	OLSR	3
Collision detection	AODV	0.915
Overtake	OLSR	0.915

values. The AODV protocol was found to be the best one for the distribution of nodes, when compared to the OLSR. Therefore, 0.915 GHz with the AODV protocol can be used to implement this scenario in real time. In the overtake scenario, the PDR was found to be 100% for all the frequencies with the OLSR protocol. The average delay of the OLSR protocol is the minimum at 0.915 GHz. The results obtained clearly show that the frequency of 0.915 GHz with the OLSR protocol was found to be the best one with 100% PDR and the minimum average delay for implementing this scenario. Table 1 shows the suitable protocol and frequency for all the scenarios.

7 Conclusions and Future Scope

Serious problems in traffic management like congestion at the traffic junctions, blocking the emergency vehicle, accidents, and consequences of overtaking are considered. The solution for the above-mentioned problems is proposed by simulating five scenarios using a C++ based simulator. The performance evaluation environment for all the scenarios with the OLSR and AODV routing protocols has been created. To evaluate the performance, video and constant bit rate data have been transmitted. The performance metrics like PDR and average delay are calculated, analyzed and graphs are plotted for the different frequencies. As a result of this analysis, the suitable protocol and frequency were proposed for each scenario. The proposed system can be enhanced in the future, when drivers can get their destination routes before starting their journey by sending a GSM based SMS to the central node. Apart from the topology-based protocols other category of protocols can also be used in such situations to see the efficacy of each one of them. Entertainments like multimedia and internet facilities can be enjoyed inside the vehicle. Weather conditions and road condition information can also be sent.

Acknowledgements The authors would like to thank the support extended by Kumaran Systems, Chennai, India for taking part in the discussions and for providing various suggestions during this research work. The authors would also like to thank the unknown reviewers who have reviewed the chapter.

References

1. Wang, Q., Hu, J., Wang, Y., & Zhang, Y. (2009). A simulation study for the communication subsystem of wireless traffic information service system. In *Information, Communication and Signal Processing, ICICS, 7th International Conference, Macau*, pp. 1–6.
2. Hu, L., Li, H., Xu, X., & Li, J. (2011). An intelligent vehicle monitoring system based on internet of things. In *Computational Intelligence and Security (CIS), 7th International Conference, Hainan*, pp. 231–233.
3. Ashtankar, P. P., Dorle, S. S., Chakole, M. B., & Keskar, A. G. (2009). Approach to avoid collision between two vehicles in intelligent transportation system. In *Emerging Trends in Engineering and Technology (ICETET), 2nd International Conference, Nagpur*, pp. 598–603.
4. Kumar, N., Lourenço, N., Terra, D., & Alves, L. N. (2012). Visible Light Communications in intelligent transportation systems. In *Intelligent Vehicles Symposium (IV)*, IEEE, Alcala de Henares, pp. 748–753.
5. Khanafer, M., Guennoun, M., & Mouftah, H. T. (2009). WSN architectures for intelligent transportation systems. In *New Technologies, Mobility and Security (NTMS), 3rd International Conference*, Cairo, pp. 1–8.
6. Nafi, N. S., & Khan, J. Y. (2012). A VANET based intelligent road traffic signalling system. In *Telecommunication Networks and Applications Conference (ATNAC)*, Australasian, Brisbane, QLD, pp. 1–6.
7. Abbas, M. K., Karsiti, M. N., Napiah, M., & Samir, B. B. (2011). Traffic light control using VANET system architecture. *National Postgraduate Conference* (*NPC*), Kuala Lumpur, pp. 1–6.
8. Bi, Y., Zhao, H., & Shen, X. (S). (2009). A directional broadcast protocol for emergency message exchange in inter-vehicle communications. *Communications, ICC'09, IEEE International Conference*, Dresden, pp. 1–5.
9. Mandal, K., Sen, A., Chakraborty, A., Roy, S., Batabyal, S., & Bandyopadhyay, S. (2011). Road traffic congestion monitoring and measurement using active RFID and GSM technology, *Intelligent Transportation Systems (ITSC), 14th International IEEE Conference*, Washington, DC, pp. 1375–1379.
10. Li, F., & Wei, Y. (2008). A real-time vehicle routing system for RFID tagged goods transportation, service operations and logistics and informatics. *IEEE/SOLI, IEEE International Conference* (Vol. 2), Beijing, pp. 2892–2897.
11. SAPS. *An analysis of the national crime statistics*. Retrieved June 22, 2015, from http://www.saps.gov.za/about/stratframework/annual_report/2013_2014/crime_stat/report_2014_part1.pdf
12. Kaur, M., Sandhu, M., Mohan, N., & Sandhu, P. S. (2011). RFID technology principles, advantages, limitations &its applications. *International Journal of Computer and Electrical Engineering, 3*, 151–157.
13. Mihret, E. T., & Ababu, K. (2019). Implementation of VANET communications: The convergence of UAV system with LTE/4G and WAVE technologies. *International Research Journal of Advanced Engineering and Science, 4*(1), 2455–9024.
14. Tilahun, E. (2018). A performance optimizing of VANET communications: The convergence of UAV system with LTE/4G and WAVE technologies. *Global Scientific Journals, 6*(12), 32–39.
15. Chiti, F., Fantacci, R., Giuli, D., Paganelli, F., & Rigazzi, G. (2017). Communications protocol design for 5G vehicular networks. In W. Xiang, K. Zheng, & X. Shen (Eds.), *5G mobile communications*. Cham: Springer.
16. Turcanua, J., Pierpaolo, S., Andrea, B., Francesca, C., & Thomas, E. (2020). A multi-hop broadcast wave approach for floating car data collection in vehicular networks. *Vehicular Communications, 24*, 100232.

Part II
Connected Vehicles

Autonomous Driving Cars: Decision-Making

Prabhu Ramanathan and Kartik

1 Introduction

Full automation is a reality in several sectors in today's world. In manufacturing industries, the everyday task is carried out by an automated process. For example, robotic arms are used on the assembly line and commercial airlines these days are almost automatic; a pilot is simply required for monitoring.

Approximately 1.35 million lives are lost due to road accidents each year [1] and most accidents are due to human error. Autonomous cars can enhance the safety and productivity of the transportation system. This requires advances in many fields of vehicle autonomy, which includes vehicle design to control, perception, planning, coordination, and human interaction.

Autonomous cars are the ones in which human drivers are not really required to take control of the safe operation of the car [2]. Autonomous cars are also called "Driverless car," "Self-driving car," "Robot car." Autonomous driving cars use various kinds of technologies, a variety of sensors are used to understand their surrounding such as radar, sonar, GPS, inertial measurement units [2]. This sensor information is used by an advanced control system to identify the appropriate path as well as an obstacle, read and react according to traffic rules (signs) and signals (Fig. 1).

Currently, over 33 companies are working on an autonomous and semi-autonomous car [3]. There are partially-autonomous vehicles with varying levels of automation and driving module [4].

The level of automation of a vehicle varies from a human-operated vehicle to a completely autonomous vehicle. SAE International outlines five levels of

P. Ramanathan (✉) · Kartik
Department of Engineering Science, University West, Trollhättan, Sweden
e-mail: prabhu.k.ramanathan@hv.se

© Springer Nature Switzerland AG 2021

N. Gupta et al. (eds.), *Internet of Vehicles and its Applications in Autonomous Driving*, Unmanned System Technologies, https://doi.org/10.1007/978-3-030-46335-9_3

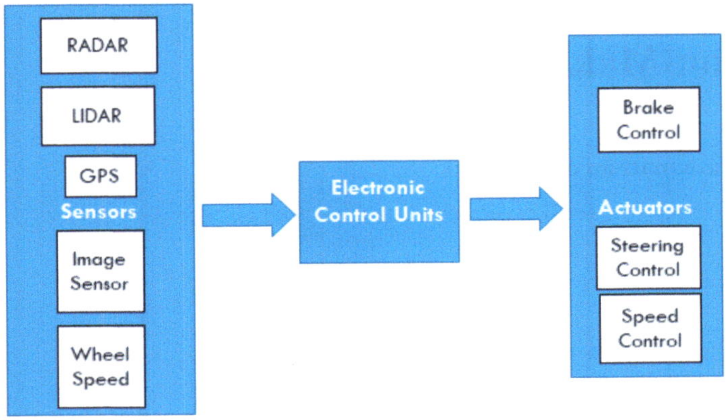

Fig. 1 Interaction of hardware and software components in autonomous system

autonomy in their J3016 document [5]. There have been advancements in the level of automation in cars with systems that support the driver. For example, to maintain a constant speed of the vehicle or follow a path. Till level 2, the driver monitoring is required at all times; in levels 3 and 4, driver monitoring is required only dependent on the situation, and level 5 is fully autonomous under all circumstances.

Levels of autonomous driving vehicles according to SAE International are,

- *Level 0*: *No automation*, [6] all major controls of the car are by the human, autonomous system may issue warnings.
- *Level 1* (*hands-on*): *Driver Assistance*, [6] both driver and autonomous system are responsible for the control of a car. But still, the driver is responsible for control. The autonomous system has control over brakes (automatic braking), cruise control (steering is controlled by a driver, the system controls the speed) lane keep, and parking assistance.
- *Level 2* (*hands-off*): *Partly automated driving*, [6] the autonomous vehicle takes care of car control. However, a driver is required for monitoring.
- *Level 3* (*eyes off*): *Conditionally automated driving*, [6] driver's complete attention is not required to monitor or control, he can text while driving as well or might even attend a call.
- *Level 4* (*mind off*): *Highly automated driving*, [6] no attention of the driver is required; he can watch a movie or eat a meal while driving.
- *Level 5* (*steering well optional*): *Full automation*, [6] no driver is required. Example: A robotic taxi.

Benefits include reduced cost, reduction in the traffic collision, provide enhanced mobility for children, elderly, disable, increases the efficiency of a vehicle, and significantly reduces to the parking space problem.

The problem includes safety, technology, hackers, and concerns about the resulting loss of driving-related jobs in the road transport industry, decision-making.

Issues were also found with driver training, mental model development, mode confusion, unexpected mode interactions, situation awareness, and susceptibility to distraction [7].

Inspired by the possibility of a future where transportation becomes a utility, academic, and industry communities have started to address the science and engineering of autonomy and significant work [8].

This literature study is about various methodologies of decision-making and planning for autonomous vehicles from the published articles.

2 Related Work

The decision-making module is one of the most important components of autonomous vehicles, connecting the environment [9] perception and vehicle control. Thus, numerous research works are performed to handle autonomous driving decision-making problems in the last decade [9].

2.1 Vehicle Dynamics and Control

At low speeds, a kinematic model of the car can be used for control and proportional–integral–derivative (PID) controller (PID controller is a control loop feedback mechanism, it calculates error value continuously that is the difference between desired value and measured value and applies correction [10]) or feedback linearization (feedback linearization is used in controlling nonlinear systems, it involves the transformation of a nonlinear system into an [11] equivalent linear system through controlled input and change of variables) can be used to track path. At high speeds (performing aggressive maneuvers), a full dynamic model of the car can be used to control. Model predictive control (MPC—It is used for controlling a process along with a satisfying set of constraints [12]) or feedback-feedforward (feedforward is a control signal given from the control system to its external environment [13]) control can be used to track path [8].

Control methods depend on a model of the vehicle [8], so the model of the vehicle is to be identified first. For system identification, both optimization-based and learning-based techniques exist [14]. The technique to be chosen will depend on the amount and type of data [8] available. Since the conditions of the road and vehicle are different at the time, online model identification [15] and lifelong system identification will improve the performance of autonomous vehicles. Tools from machine learning are used to create models from large amounts of data collected [8].

2.2 Collaborative Autonomy

Collaborative autonomy is a basic principle supporting collaborative intelligence through which one contributes by applying unique skill and leadership autonomy in problem-solving [16]. There are three types of collaborative autonomy:

1. Series autonomy, here human commands the vehicle to execute a function; this is similar to most of the autonomous driving systems to date [8].
2. Interleaved autonomy, here the human driver and the autonomous system take turns operating the vehicle [8].
3. Parallel autonomy (also referred to as shared control), in which the autonomous system functions as a savior in the background to ensure safety while the human driver is driving the vehicle. Whether drivers are distracted or not, a parallel autonomy provides additional safety [8].

2.3 Path Planning for Autonomous Vehicles Using Model Predictive Control

Path designing for autonomous vehicles in dynamic environments is vital, however, but challenging, due to the constraints of vehicle dynamics and the existence of surrounding vehicles. Typical trajectories of vehicles involve completely different modes of maneuvers, as well as lane-keeping, lane change, and intersection crossing. There exists previous arts exploitation the rule-based high-level deciding approaches to determine the mode shift. Instead of exploiting specific rules, authors propose a unified path designing approach exploitation model predictive control (MPC) that mechanically decides the mode of maneuvers. To ensure safety, they have developed a kind of constraints in model predictive control to enforce the collision avoidance between the autonomous vehicles and surrounding vehicles. To achieve comfortable and natural maneuvers, the lane-associated potential field in the objective function of the model predictive control was included [17] (Fig. 2).

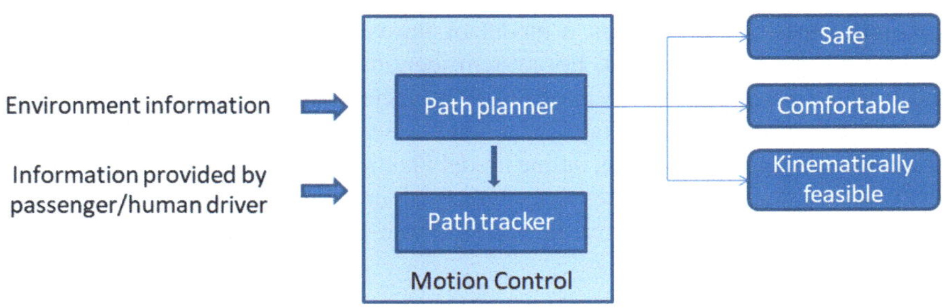

Fig. 2 Motion control strategy of autonomous vehicle using MPC. Modified from [17]

The proposed planner automatically decides the mode of maneuvers under a unified optimization framework. A convex relaxation approach (an approximation of complex problem to a nearby problem that is easier to solve [18]) is used to determine the lane change and lane-keeping maneuvers. They have a collision avoidance condition to ensure the safety of the autonomous vehicle [19]. The simulation has been conducted for different situations, including lane change, lane-keeping, and intersection crossing. Results have shown that the proposed path planner is effective in generating a safe [17], kinematically feasible and comfortable path for autonomous vehicles. Future work of the author will focus on the implementation of the path planner in real experiments [17] and also the prediction of surrounding vehicle motion.

2.4 *Empirical Decision-Making System for Autonomous Vehicles*

The major cause of an accident between an autonomous car and the human-driven car is the misunderstanding between autonomous driving systems and human drivers. To solve this problem the authors in [20] have proposed a humanlike driving system, to give autonomous vehicles the ability to make decisions like a human. In the method, a convolutional neural network (CNN) model is used to detect, recognize, and abstract the information from the road scene, which is captured by the onboard sensors. They built a perception system with only the depth information, rather than the unstable RGB data. This system includes two parts, that is, a road scene perception method and an empirical decision-making network. The difference with the existing approaches is that it can imitate human drivers' social intelligence, which can better adapt the self-driving vehicles to the real-life road conditions. In addition, they implement an effective training scheme to improve the quality and speed of data collection and the time-consuming manual labeling process. Also, they can find a feasible approach to analyze the possible influence factors in [21] the decision-making process, which can help in the testing and validating of autonomous driving systems. The experimental results prove that the proposed method is efficient [20]. Future work of authors includes developing a more efficient optimization method to decrease the time cost of the training process. Also, some information may be extracted from RGB data, even when it is incomplete [22] and unstable, which can be a useful supplement to this method (Fig. 3).

2.5 *Real-Time Decision-Making for Autonomous City Vehicle*

The authors in [23] paper address the topic of real-time decision-making by autonomous city vehicles. The paper presents the vehicle decision-making and

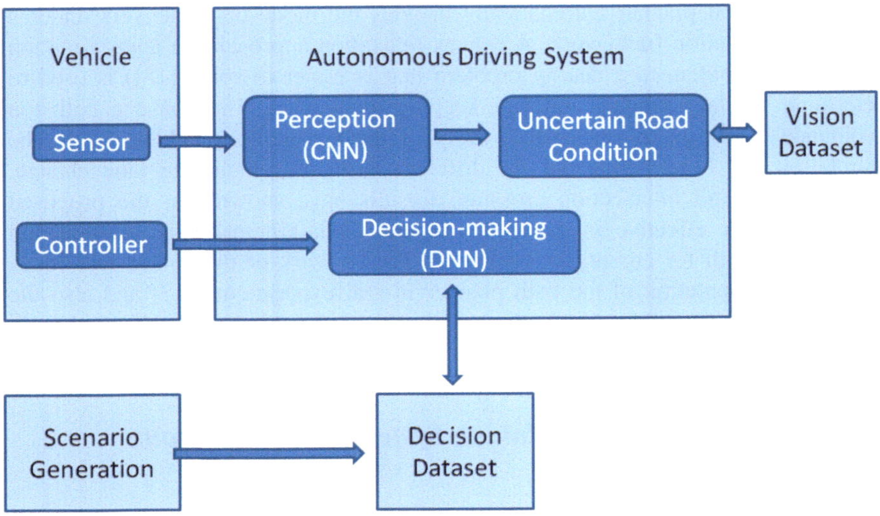

Fig. 3 System framework. Modified from [20]

control system architecture, explains the subcomponents which are relevant for decision-making (World Model and Driving Maneuver subsystem), and presents the [23] decision-making process.

The solutions developed for simplified traffic conditions, the paper has presented the vehicle control system architecture and addressed in more detail the three main subsystems relevant for decision-making: the World Model, the Driving Maneuver subsystem, and the Real-Time Decision-Making subsystem. The main role of the World Model is to provide a Real-Time Decision-Making subsystem with exact and update information about the vehicle's traffic environment. The Driving Maneuver subsystem is the output results of decision-making. Driving maneuvers are closed-loop control algorithms, each capable of maneuvering the autonomous vehicle in [24] a specific traffic situation. Each driving maneuver is modeled based on a common finite automaton structure, which is capable of modeling complex driving maneuvers while enabling the decision-making module to start and stop their execution. Furthermore, the developed finite automaton model enables the decomposition of complex driving maneuver control algorithms into subtasks with manageable complexity, and their implementation using closed-loop control methods in the Run states (Fig. 4).

The purpose of the Real-Time Decision-Making subsystem is to make driving decisions based on information from the World Model and to activate the execution of the most appropriate driving [25] maneuver. The task of decision-making is decomposed into two consecutive stages [23]. The first, safety-crucial decision-making stage determines which driving maneuvers are feasible, that is, safe to perform, conforming to traffic rules, and in line with the path planner indication.

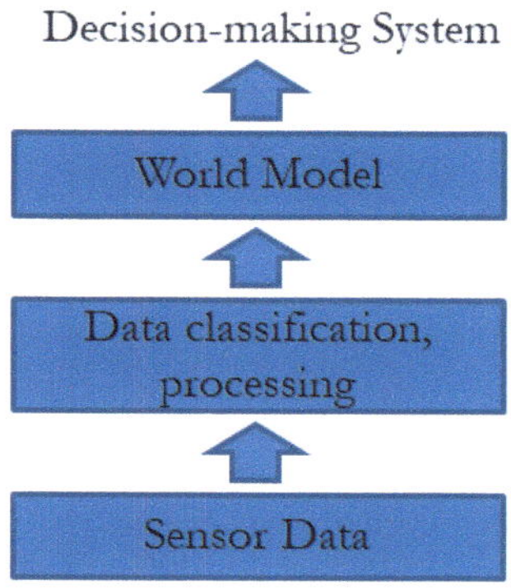

Fig. 4 World model input, and output. Modified from [23]

The second decision-making stage decides the best execution alternative among the feasible driving maneuvers. The main focus of this stage is to improve comfort and efficiency.

2.6 Intention-Aware Autonomous Driving Decision-Making

In paper [9] an intention-aware decision-making algorithm for the autonomous driving system is proposed. This is to solve the problem of uncertain situations like an uncontrolled intersection. Hidden Markov model (HMM—the system model in which sequence possible of with uncontrollable states [26]) was built to predict uncertain intentions and partially observable Markov decision process (POMDP is generalized Markov decision process [27, 28]), a general decision-making process was modeled and then use an approximate approach to solve this complex problem since solving by POMDP is difficult. Finally, PreScan software (A software tool which monitors the vehicle's surroundings using the information given by sensors [29]) and a driving simulator (driving simulators are used to teach human drivers and they are used for research purpose in the area of human factors and medical research, to monitor drivers behavior, attention, and performance [30]) were used. A driving simulator was used to emulate the social interaction process. The experimental results show that autonomous vehicles with this approach can pass through uncontrolled [9] intersections more safely and efficiently than using the strategy without considering human-driven vehicles' driving intentions.

3 Summary

This study is an outline of current methodologies made in the field of planning and decision-making for autonomous vehicles. Table 1 summarizes the important articles published in this domain. While this field has gained ground in the past years, many questions remain unanswered. For autonomous systems to operate

Table 1 Summary of the articles

S. No.	First author	Year	Topic of discussion	Comment
1	Rajesh Rajamani	2012	Vehicle dynamics and control	Models and feedback mechanism to control the vehicle at different speeds and to obtain corresponding path of the vehicle
2	Rubin MA	2014	Collaborative autonomy	Different types of collaborative autonomy and who will be responsible for the leadership of decision-making (autonomous vs human driver)
3	Chang Liu	2017	Path planning for autonomous vehicles using MPC	Path planning for autonomous vehicle (lane-keeping, lane change, and intersection crossing) using MPC
4	Liangzhi Li	2018	Empirical decision-making system for autonomous vehicles	Analysis of the external environment and decision-making like human using CNN/ConvNet
5	Andrei Furda	2010	Real-time decision making for Autonomous City vehicle	Vehicle decision-making and control system architecture (world model and driving maneuver subsystem)
6	Weilong Song	2016	Intention-aware autonomous driving decision-making	Hidden Markov model and partially observable Markov decision process can be used to solve uncertain situations like an uncontrolled intersection

in complex, dynamic, and interactive environments require artificial intelligence. Autonomous systems still need to reach the human-level in decision-making, planning, and perception. And with current advancement is not pleasing. In future autonomous vehicles will provide on-demand transportation to anyone, anytime, anywhere. To achieve this vision, further improvements are required in the field of stochastic routing, online performance, and quality of service. If we can beat these challenges, autonomous vehicles will have an enormously valuable effect on our lives.

References

1. WHO. [Online]. Retrieved December 7, 2018, from https://www.who.int/news-room/fact-sheets/detail/road-traffic-injuries
2. Education Ecosystem. [Online]. Retrieved from https://www.education-ecosystem.com/guides/x/self-driving-cars/history

3. Cheung, K. Algorithm-X Lab. [Online]. Retrieved March 28, 2019, from https://algorithmxlab.com/blog/worlds-top-33-companies-working-on-self-driving-cars/
4. Wikipedia. [Online]. Retrieved from https://en.wikipedia.org/wiki/Self-driving_car
5. SAE International. [Online]. Retrieved June 15, 2018, from https://www.sae.org/standards/content/j3016_201806/
6. [Online]. Retrieved from https://www.cnet.com/roadshow/news/self-driving-car-guide-autonomous-explanation/
7. Luzuriaga, M., Kunze, O., & Heras, A. (2019). Hurting others vs hurting myself, a dilemma for our autonomous vehicle. *Review of Behavioral Economics, 7*(1), 1–30.
8. Schwarting, W., Alonso-Mora, J., & Rus, D. (2018). Planning and decision-making for autonomous vehicles. *Annual Review of Control, Robotics, and Autonomous System, 1*, 187–210.
9. Song, W., Xiong, G., & Chen, H. (2016). Intention-aware autonomous driving decision-making in an uncontrolled intersection. *Mathematical Problems in Engineering, 2016*, 1–15.
10. Wikipedia. (2019). [Online]. Retrieved from https://en.wikipedia.org/wiki/PID_controller
11. Wikipedia. (2019). [Online]. Retrieved from https://en.wikipedia.org/wiki/Feedback_linearization
12. Wikipedia. (2019). [Online]. Retrieved from https://en.wikipedia.org/wiki/Model_predictive_control
13. Wikipedia. (2019). [Online]. Retrieved from https://en.wikipedia.org/wiki/Feed_forward_(control)
14. Nelles, O. (2001). *Nonlinear system identification: From classical approaches to neural networks and fuzzy models*. New York: Springer.
15. Seegmiller, N., Rogers-Marcovitz, F., Miller, G., & Kelly, A. (2013). Vehicle model identification by integrated prediction error minimization. *International Journal of Robotics Research, 32*(8), 912–931.
16. Gill, Z. [Online]. Retrieved from http://collaborative-intelligence.org/autonomy.html
17. Liu, C., Lee, S., Varnhagen, S., & Tseng, H. E. (2017). Path planning for autonomous vehicles using model predictive control, *28th IEEE Intelligent Vehicles Symposium*, CA.
18. Wikipedia. [Online]. Retrieved March 6, 2018, from https://en.wikipedia.org/wiki/Relaxation_(approximation)
19. Ventura, J., Ciarcia, M., Romano, M., & Walter, U. (2016). An inverse dynamics-based trajectory planner for autonomous docking to a tumbling target. *AIAA Guidance, Navigation, and Control Conference*, San Diego, CA.
20. Liangzhi, L., Kaoru, O., & Mianxiong, D. (2018). Human-like driving: Empirical decision-making system for autonomous vehicles. *IEEE Transactions on Vehicular Technology, 67*(8), 6814–6823.
21. Zimmermann, H.-J. (2001). *Fuzzy set theory and its applications* (4th ed.). Dordrecht: Springer.
22. Russ, J. C. (2018). *The image processing handbook* (3rd ed.). Boca Raton: CRC Press.
23. Ljubo Vlacic, A. F. (2010). Real-time decision making for autonomous city vehicles. *Journal of Robotics and Mechatronics, 22*(6), 694–701.
24. Andrei Furda, L. V. (2011). Enabling safe autonomous driving in real-world city traffic using multiple criteria decision making. *IEEE Intelligent Transportation Systems Magazine, 3*(1), 4–17.
25. Barjonet, P.-E. (2001). *Traffic psychology today*. Boston: Kluwer Academic Publishers.
26. Wikipedia. (2019). [Online]. Retrieved from https://en.wikipedia.org/wiki/Hidden_Markov_model
27. Wikipedia. (2019). [Online]. Retrieved from https://en.wikipedia.org/wiki/Markov_decision_process
28. Wikipedia. (2019). [Online]. Retrieved from https://en.wikipedia.org/wiki/Partially_observable_Markov_decision_process
29. Advanced sim tech. [Online]. Retrieved from http://www.advancedsimtech.com/software/prescan/
30. Wikipedia. (2019). [Online]. Retrieved from https://en.wikipedia.org/wiki/Driving_simulator

IEEE 802.11ah for Internet of Vehicles: Design Issues and Challenges

Badarla Sri Pavan, Miriyala Mahesh, and V. P. Harigovindan

1 Introduction

The Internet of Things (IoT) is a rising and predominant paradigm shift that anticipates the near future of the global world in the era of future Internet through interconnected physical objects. The recent advancements in wireless technologies aid the emergence of Internet to be ubiquitous in the real world to share the huge amount of information through the various connected objects for different application scenarios [24]. The term IoT is first coined by Kevin Ashton in 1999. It is forecasted that the number of devices estimated to be one trillion by 2035 [20]. IoT is a platform to interconnect the things and activate the Machine to Machine (M2M) communication to share the information among the heterogeneous objects by excluding the human intervention [10]. IoT enables the connected environment that will affect the various aspects of human life and industrial applications based on the nature of services and applications.

In IoT, things are connected to capture and share the information that includes computers, electronic devices, sensors, actuators, people, vehicles, trees, etc. [2]. It aims to enable the future Internet to be ubiquitous and pervasive. Moreover, the advancements in wireless technologies facilitate the easy access and interaction among the devices which tends to increase the expansion of various applications and helps to share the huge amount of data. The main asset for the interconnection and networking of smart things is Big Data that capture and deliver valuable information [38]. This provides various services to the people, companies, public and private administrations. Hence, IoT concept grabs high attention in the various application scenarios for different domains such as smart home, smart industry, smart grids,

B. S. Pavan (✉) · M. Mahesh · V. P. Harigovindan
Department of Electronics and Communication Engineering, National Institute of Technology Puducherry, Karaikal, Union Territory of Puducherry, India
e-mail: sripavan.rvce@gmail.com; hari@nitpy.ac.in

© Springer Nature Switzerland AG 2021

N. Gupta et al. (eds.), *Internet of Vehicles and its Applications in Autonomous Driving*, Unmanned System Technologies, https://doi.org/10.1007/978-3-030-46335-9_4

smart transportation systems, smart health-care, etc. [1]. However, accomplishing the requirements of various applications is a big challenge. This complexity can lead to proliferation of various proposals for the practical implementation of the IoT systems. The support of various applications makes the IoT as a cross-platform for things to communicate at any time, anywhere, and with anything [36].

IoT targeted to provide association among various things that use a wide range of heterogeneous technologies to support various utilities. This can improve the different aspects of the country such as economy and social life. IoT desires to have multiple features that include identification, sensing, communication, hardware and software technologies for various platforms to support different applications. Among several communication technologies, there are two types of low power IoT wireless communication technologies. Firstly, Wireless Personal Area Networks (WPAN) which is intended for wireless transfer of data over short transmission ranges. Secondly, Low Power Wide Area Networks (LPWAN) developed for large coverage area but it is limited to lower throughput. However, WPAN networks which are Bluetooth and ZigBee have short range of transmission and LPWAN networks which are SigFox and Long Range Wide Area (LoRa) suffers from very low throughput. Hence, these technologies can be applied to very few IoT scenarios [34]. Therefore, to support the diverse applications of IoT, wireless communication technologies have to incorporate the features such as high throughput and large coverage.

Due to the popularity of Wireless Local Area Networks (WLAN), IEEE 802.11 brings various amendments to the legacy 802.11 standard. In dense networks, IEEE 802.11 suffers from severe contention during the channel access procedure. Hence, IEEE 802.11ah have been announced by the IEEE 802 task group with ample features to support various IoT applications [5]. Along with existed Wireless-Fidelity (Wi-Fi) standards, it features various Modulation and Coding Schemes (MCS) to provide the extended coverage, optimal data rate, and ubiquitous connectivity. The standard allows to support data rates from 0.15 to 78 Mbps in the range of 100 m to 1 km [31]. Thus, the IEEE 802.11ah standard can achieve higher transmission ranges in comparison to the WPAN and LPWAN technologies. Additionally, it provides several enhancements to the Physical (PHY) and Medium Access Control (MAC) layer [18]. It includes various innovative mechanisms in the MAC layer such as hierarchical Association Identifier (AID) structure to support large number of devices, minimization of overhead by including short headers, Restricted Access Window (RAW) to mitigate the contention among the densely deployed nodes, Traffic Indication Map (TIM), and Target Wakeup Time (TWT) for energy efficiency [41].

From the literature, it is observed that the number of vehicles in the world is around one billion and predicted to be two billion by 2035 [9]. In recent years, the world has evidenced that the explosive growth of number of vehicles in the transportation systems for traveling purposes. Among several IoT applications, the most important aspect of Intelligent transportation systems (ITS) is to assure the safe delivery of passengers. Due to advancements and automation, the high mobility of vehicles makes the traffic management very complex. It leads to traffic congestion, fatalities due to accidents, and also waste lot of time in the traffic which severely affects the economic growth of the country [3]. Moreover, technological advancements in the automation industry causes the increase in usage of number of vehicles, as a result the transportation systems are facing many technical challenges. Many issues such as accuracy, delay sensitivity, seamless connectivity, and data privacy have to be solved in transportation systems. Besides, based on the low cost hardware, a vehicle generally contains a limited processing and storage to support several applications. Hence, it is indeed to provide an efficient transportation system with improved traffic management by intimating the drivers regarding the optimum routes and safety messages.

To bring the advancements in the transportation systems, Internet of Vehicles (IoV) has been introduced as a promising branch in the new era of evolution of IoT. It is a unified network that aims to intelligently manage the traffic and dynamically control the vehicle in ITS using IoT technology. In another way, it is a platform that connects various vehicles, humans, things on the roadside infrastructure through the IoT enabled network [30]. The technological advancements open a gate way to the IoV with intelligent devices (vehicles) that are endowed with embedded processors and wireless technologies. Several wireless technologies play an essential role in making the transportation systems smarter with connected cars, vehicle to vehicle communication, and with surroundings. In the USA, many online devices, including vehicles, are equipped with security chips to get the identity of every device on the Internet [17]. In Delhi, all the vehicles, including registered autos, government buses, are built with Global Positioning System (GPS) and Wi-Fi.

Figure 1 shows the scenario of IoV by enabling IoT network. It is to capture and share the valuable information among the vehicles, roads, and surrounding objects. It features acquisition, filtering, processing, securing, managing, and

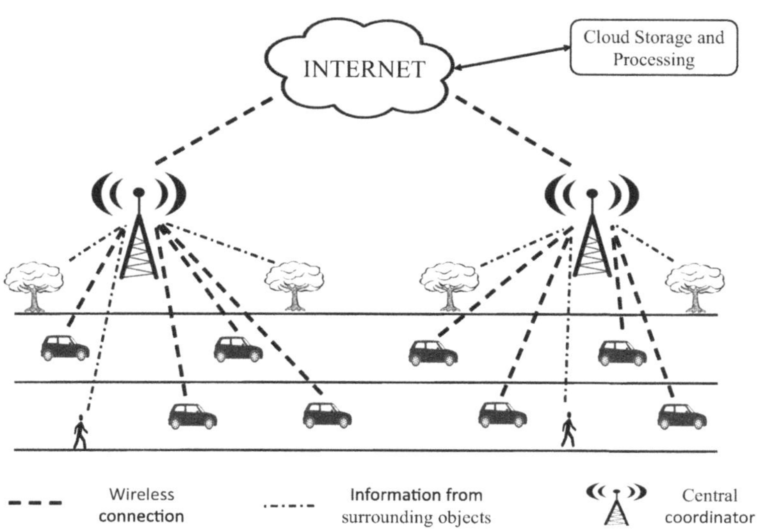

Fig. 1 IoV with IoT enabled network scenario

communication of information to different platforms [16]. The available information is used to guide the vehicles and provide various application services to make the transportation systems dynamic and economical.

In IoV network, vehicles share the information related to vehicle location, speed, and traffic conditions that aid safe traffic management, lower traffic congestion, and reduce road accidents. Recently, cloud storage is integrated with the IoV that is capable of storing and processing large amount of data and sharing the information with the connected vehicles, humans, and drivers. The stored data can be processed and computed with the cloud computing [39]. With interconnection and interoperability, intelligent devices provide an accident-free and comfortable environment. Besides, because of the ubiquitous Internet access, there can be a good chance for emerging new products and services [13].

IoV is a very complex system with different networking components and supports various applications. To make the IoV applications practical, various networking technologies have to be integrated with necessary features. To manage the massive number of heterogeneous devices in IoV and to make the feasible communication among them is a critical task. Hence, to support such dense IoV networks, MAC layer plays a key role in various aspects. A lot of research is going on to design the MAC layer protocols for IoV networks [26]. Despite the popularity of various WLAN standards, IEEE 802.11 which is intended to provide the PHY and MAC layer protocols to support numerous applications. Various amendments to the legacy IEEE 802.11 standard have been introduced subsequently. Among them, IEEE 802.11p/WAVE has attracted more attention in vehicular environments. However, it suffers because of limitations such as (1) Crowded spectrum around 5 GHz frequency band (2) High mobility of vehicles (3) Low data rate (4) High contention when large number of devices supposed to access the channel [35]. Thus, for the IoV scenario the standard is not a suitable candidate due to its limitations. Recently, the IEEE 802 LAN/MAN committee has introduced the IEEE 802.11ah for dense networks. The standard provides numerous advantages [4, 27] such as (1) Operates in sub-1 GHz bands (less crowded spectrum) (2) Long transmission range of 1 km (3) High data rates ranging from 0.15 Kbps to 78 Mbps (4) Reduces the contention among the devices by MAC layer mechanism known as RAW (5) Energy efficient mechanisms. Hence, IEEE 802.11ah is the perfect candidate to support dense networks for the IoV scenario.

Different sections of the chapter are organized as follows. Section 2 provides the insight view of architecture, characteristics, and benefits of IoV networks. Various features and enhancements of IEEE 802.11ah PHY and MAC layers are discussed in Sect. 3. The key design issues and challenges of IEEE 802.11ah in IoV networks are described in Sect. 4. Finally, the chapter is concluded with Sect. 5.

2 Internet of Vehicles

Nowadays, the living standards of people determine the social economy of the country. Due to modernization and automation, there is a drastic growth in the quantity of number of vehicles for transportation purposes. However, due to drowsy driving, blocked visibility of vehicles, high speeding leads to road accidents and human life is unsafe while traveling. From the past reports, most of the accidents occurred due to the errors in human driving and mis-intuitions. Hence, to minimize the fatalities due to accidents and to make the transportation systems efficient, IoV has been introduced as a promising stream of IoT. The concept of IoV is a new area for industry, academia, and researchers. It refers to the heterogeneous interconnection of huge number of vehicles that share information through the Internet. This concept is getting attention because of the increase in the number of vehicles along with mobile equipment such as smartphones, wearable smart sensors, and many more. Many wearable sensors such as motion sensor, smart glasses, ECG sensor, alcohol detecting sensors, smart watch, etc. can be used to monitor the condition of driver and passengers. Similarly many other sensors such as headlight range sensor, mirror sensor, transmission sensor, fuel level sensor, smoke sensor, vehicle speed sensor, GPS sensor, etc. are used to monitor the vehicles condition. In the context of IoV, vehicles are equipped with multiple sensors, robust computational points, and connectivity to the Internet based on modern technologies [7].

IoV enables to collect and distribute the information related to the vehicle's navigation, road condition, environmental conditions, and surroundings. Moreover, it is equipped with versatile processing, computing, distributing, security, and privacy of data to be provided to the different information platforms such as Internet systems. Depending on the collected information the central coordinator can guide the vehicles to move in the proper direction, intelligent traffic management and provide various multimedia and Internet applications. In IoV networks, vehicles are smart and intelligent with various advanced features. The sensors gather the vehicular movement, position, and driving condition and forward it to the other vehicles and Internet with an equipped embedded platform. Cloud storage plays a major role in IoV networks. The collected huge amount of Big Data among the vehicles is stored in cloud storage. Cloud should have powerful storage and computing resources. Besides, IoV scenario have two technological goals (1) Networking among the vehicles (2) Intelligence among the vehicles. The aim of networking is to integrate various objects such as people, vehicles, things, networks, and the environment. The intelligent network is created with efficient processing and communication capabilities that support various services for large-scale cities and countries. There are many issues to be considered in IoV due to applications and service requirements such as the dynamic nature of vehicles movement, connectivity, and privacy of data, etc.

2.1 Architecture of IoV

The architecture of IoV is categorized into three main sections. The first section comprises of sensors placed in and around the vehicle which collects the information from the specific events and environments such as the vehicle condition, movement, and environmental conditions with others. The second section consists of devices with a wireless connection such as vehicle to central coordinator, vehicle to vehicle, vehicle to sensor, and vehicle to environment conditions. It provides ubiquitous connectivity to the available networks by Wi-Fi, GSM, LTE, Bluetooth, and IEEE 802.15.4 with others. The third section consists of different tools, cloud for storage, infrastructure, and Big Data for the processing. It is accountable for storage, investigation, processing, and decision making in dangerous situations (such as risky road conditions and traffic congestion). The aim is to choose the correct decisions based on the data collected from the various systems and communication technologies [12].

From the services-based approach, IoV is organized into seven layers as shown in Fig. 2 [9]. The responsibilities of each layer are discussed as follows. The user interface layer is to provide direct communication with the driver using the management interface to synchronize the driver notifications and choose the best decision to avoid the collision risks with other vehicles. For example, if there is a probability of collision with a vehicle foremost, an indication of lights on the dashboard of the car can be active while sound will horn to make the driver attentive.

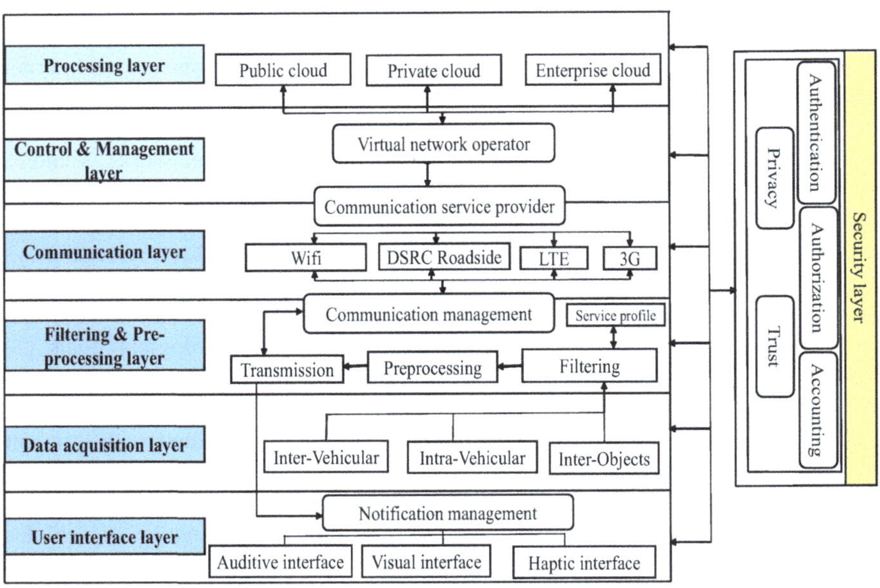

Fig. 2 IoV layer architecture

This layer provides the interaction between the driver and vehicle. The way of communication between the driver and vehicle can be of three ways. In an auditive interface, the messages are available to the driver by sound indication. The driver can listen the sound-based notification and react accordingly. In a visual interface, the driver can visualize the indications through graphical icons. Differently, haptic interface is a kind of system which enables the human to interact with an intelligent computer uses bodily sensations and movements. The data acquisition layer gathers the information from the various smart devices such as sensors within the vehicle, navigation system, inter-vehicle communication, various other equipped sensors, traffic lights, and signals among other vehicles placed on the roads. The received information from various sources is manipulated by the data filtering and pre-processing layer. It removes unwanted data and mitigates network traffic. Here, traffic decisions are based on the service profile created for vehicles based on active/subscribed services.

The communication layer is responsible to choose the efficient network to transmit the information based on various parameters such as similar data, level of the Quality of Service (QoS) in available networks, privacy, and security. For example, if we want to use a Wi-Fi network, it chooses the best service provider which depends on many factors such as communication requirements, profile of the vehicle, network quality, transaction cost with others to maintain efficient communication. The control and management layer functionality are to manage the various service providers in the IoV environment. In this layer, there is a possibility to apply various policies and functions for better management of received information. The processing layer is responsible for the processing of a huge amount of data in connection with the cloud computing storage infrastructures. This can be done locally and remotely. The service providers use the massive amount of processed data for further improvement of services, to enhance and develop the new applications. Processing layer will have the intelligence to handle the large volume of information. It possesses the capacity to process, analyze, and evaluate the large amount of information that is received from the various smart devices. To process the huge amount of data, processing infrastructure mainly consists of IoV intelligence and emerging technology such as Big Data to solve the various issues. Based on the advancements in Big Data technology, the processing layer will have the capacity to analyze, manage, and process the huge amount of information from various heterogeneous smart devices. The extraction and processing need expertize from the various areas such as science to government and consumers to enterprises [8]. Moreover, cloud in processing layer provides the Internet access to the smart devices for cloud based processes such as content searching, spectrum sharing, accessing the computing resources. The cloud contains the information related to smart vehicles, humans, traffic, pollution level, safe routes, etc. Hence the security and privacy of the information is a critical task. Accordingly, the data is organized in three parts such as public cloud, private cloud, and enterprise cloud. Among them the information in public cloud is available to all the public, whereas in the private cloud, a single organization exclusively operates the cloud infrastructure. The enterprise cloud is an intelligent environment for the business purpose which offers more security. The security layer has direct communication with all other layers. The objective of the security layer is to provide authentication, non-repudiation, integrity, confidentiality in the network and protect the data from malicious and cyber-attacks. Additionally, it administrates the authorization, authentication, and is responsible for all the activities among various IoV entities.

2.2 Characteristics and Benefits of IoV

Advanced technological aspects and seamless mobile Internet access have transformed the vehicles into new smart mobile device which allows the vehicles to be always connected. This enables to establish, access, and communicate a lot of useful information among the people, infrastructure, and other vehicles. It provides intelligent management of traffic, intelligent control of the vehicle in connection to the others. IoV consists of three major components (1) Inter-vehicular network (2) Intra-vehicular network (3) Vehicular mobile internet. IoV is complex network that allows vehicles and their surroundings to exchange information using advanced communication technologies. It enables the vehicle to be continuously allied with the Internet, which forms a network of vehicles that facilitate the information for various services such as traffic control and road safety. However, IoV is different from the ITS because it concentrates on the sharing of information among vehicles, people, and the surrounding environment. The characteristics of IoV networks are described as follows.

Highly Dynamic Network

Vehicles in the IoV network move with very high speed in comparison to the mobile devices in cellular networks. This makes the network to be highly dynamic in IoV scenario.

Varying Density of Network

Because of the high mobility and huge number of vehicles, the traffic density of the network varies dynamically.

Geographical Communication

In IoV network scenario, the communication among vehicles is a new type of communication in comparison with other networks. It addresses the geographical areas where the information need to be forwarded for safe driving applications.

Scalability of Network

Due to the more population, the scalability of the network can be large in urban places such as in high ways, at the center and starting point of the city.

Mobility Prediction

In the scenario of mobile ad-hoc networks, the devices move in a random direction, whereas in IoV, vehicles will move based on the layout and road topology. Hence, the vehicles movement can be predicted based on mobility.

Storage and Energy

In IoV scenario, vehicles are cars rather than small handheld devices. Hence, the vehicles will have sufficient energy since the vehicles are equipped with ample battery resources, computing, and storage capability.

Heterogeneous Communication Environment

In IoV networks, there are two types of communication environments. In highway scenarios, the traffic is simple and straightforward, whereas in city due to trees and buildings there is no direct line of sight communication always in the intended direction to share the data.

The benefits of IoV are discussed as follows.

Safe Driving

From the past many years, most of the people are dying because of road accidents. This is because of lack of information to the driver about the traffic conditions. IoV provides an efficient and advanced information very quickly to the drivers and passengers to safeguard and can make the decisions based on the traffic alerts such as information of the surrounding vehicle's speed and its future location to avoid the collisions with them. Also, it provides efficient routes for the drivers. This reduces the road accidents, and amount of time-wasting in the traffic which improves the productivity of the country.

Transportation Efficiency

Traffic control in the urban areas and at junctions is always the key issue in transportation systems. It depends on the management of traffic signals, information about the volume of the traffic. IoV can efficiently control the traffic by predicting the density of the number of vehicles at the intersection areas which saves lot of time for the people by mitigating the traffic jams.

Sociological Benefit

Nowadays, pollution in the atmosphere is a sever issue due to the massive number of vehicles in the global world. This results in global warming. With the help of IoV technology, the causes for global warming because of heavy traffic on roads at the junctions, number of crashes among vehicles, congestion control, and CO_2 emissions can be reduced by providing efficient traffic management.

Cloud Storage

In dense vehicular network scenarios, huge amount of valuable data is shared among the high mobility vehicles. IoV provides the efficient processing and storage of data with IoT enabled network.

3 IEEE 802.11ah: A Perfect Candidate for IoV

The rapid advancements in the IoT and M2M communication have facilitated the development of new communication technology that can operate in different frequency bands in a highly contending network. Among various existed standards, Wi-Fi is a popular communication technology because of its high throughput [6]. It supports various amendments to different networking requirements. Many amendment standards of legacy IEEE 802.11 operate around 2.4 and 5 GHz ISM frequency bands (IEEE 802.11a, IEEE 802.11b, and IEEE 802.11p). This makes the spectrum much denser and creates interference because of the huge number of devices. Moreover, in a dense network scenario, IEEE 802.11 WLAN faces a severe contention among the devices which degrades the network's performance [32].

Fortunately, the IEEE 802 standardization task group has introduced a new WLAN standard named IEEE 802.11ah for dense IoT networks which is intended to operate in sub-1 GHz frequency bands for PHY and MAC layers [15]. Hence, the problem of the crowded spectrum can be mitigated because of the usage of lower frequency bands (sub-1 GHz) which results in improved coverage area when compared with other WLAN standards. Due to technological developments and urbanization, the population in cities is increasing rapidly which causes the transportation systems to face lot of disputes in traffic management. It affects the growth of the country, causes road accidents, increased pollution, etc. The intelligent traffic in modern days shares more digital information through the technology and service-based systems.

IoV is a subset of IoT, mainly developed to improve the transportation systems to be more intelligent. Its main purpose is to gather, process, share, and analyze the information for the efficient transportation system. In IoV, many heterogeneous devices (vehicles) to be interconnected and predict the information of environment and road traffic. Here, the vehicles and environmental objects need to be interconnected and share the information through the Internet [26]. There is a requirement of new standard in IoV to support features such as the massive number of connected devices, long range, and scalability. IEEE 802.11ah accomplishes these requirements and is a reliable candidate to support IoV features.

3.1 PHY Layer

IEEE 802.11ah inherits the PHY layer of IEEE 802.11ac and allowed to operate in sub-1 GHz bands by excluding TV whitespace bands. The applicable sub-1 GHz ISM frequency bands for different countries are listed in Table 1. For example, in the USA, IEEE 802.11ah operates in the 915 MHz ISM bands to support IoT applications and to extend the range of Wi-Fi. It uses Orthogonal Frequency Division Multiplexing (OFDM) PHY transmission.

The PHY layer of OFDM operates at ten times down clocked version of the IEEE 802.11ac's PHY layer. PHY transmission uses OFDM based waveform consisting of 32 or 64 tones/sub-carriers (including tone allocated as pilots and guards) with 31.25 kHz spacing. IEEE 802.11ah PHY layer supports various MCS schemes. It defines five channel bandwidths in a range of 1–16 MHz (1, 2, 4, 8, and 16 MHz). For extended coverage, channel bandwidth 1 MHz is specifically designed with MCS 10 scheme. The standard uses the repetition coding scheme for 1 MHz channel bandwidth with Binary Phase Shift Keying (BPSK) modulation for 1/2 code rate. The supported modulation schemes are BPSK, Quadrature Phase Shift Keying (QPSK), 16, 64, and 256-Quadrature Amplitude Modulation (QAM) [32]. The standard also adopts Multiple Input Multiple Output (MIMO) and downlink multi-user MIMO. Physical (PHY) layer in IoV mainly deals with the movement of smart devices such as vehicles, human, things, road side infrastructure, etc. Consequently, various effects like multi path fading and Doppler effect will arise

Table 1 Sub-1 GHz ISM frequency bands

Country	Frequency bands
United States	902–928 MHz
South Korea	917.5–923.5 MHz
Europe	863–868 MHz
China	755–787 MHz
Japan	916.5–927.5 MHz
Singapore	866–869 MHz
	920–925 MHz

due to the variation in speed of vehicles. The corresponding information from the PHY layer will be directed to the higher layers. In recent years, the advancements in technologies mainly concentrated on the use of radio and infrared waves for the communication among the smart devices. The merits being, it offers a line of sight and broadcast communication. In IEEE 802.11ah, PHY layer standard makes the communication among the smart devices based on the radio waves.

3.2 MAC Layer

Several MAC enhancements to the legacy IEEE 802.11 have been introduced in IEEE 802.11ah standard draft. The improvements in MAC layer can support large number of devices, power save, channel access mechanism, and throughput [18, 25]. In the subsequent section, the features of the IEEE 802.11ah MAC layer are presented.

Support of Large Number of Associated Devices

In the IEEE 802.11ah, the Access Point (AP)/central coordinator supports large number of devices. Every device connected to the AP have a unique identifier, called as AID. It helps to identify the devices in the network. The legacy 802.11 standard supports maximum of 2007 devices per AP. Based on the AID structure, IEEE 802.11ah supports a maximum of 8192 associated devices for each AP. This helps in the IoV scenario to connect large number of vehicles.

The AID structure is organized in pages, blocks, sub-blocks, devices position index in sub-block. The AID has 13 bits as illustrated in Fig. 3 with a four-level

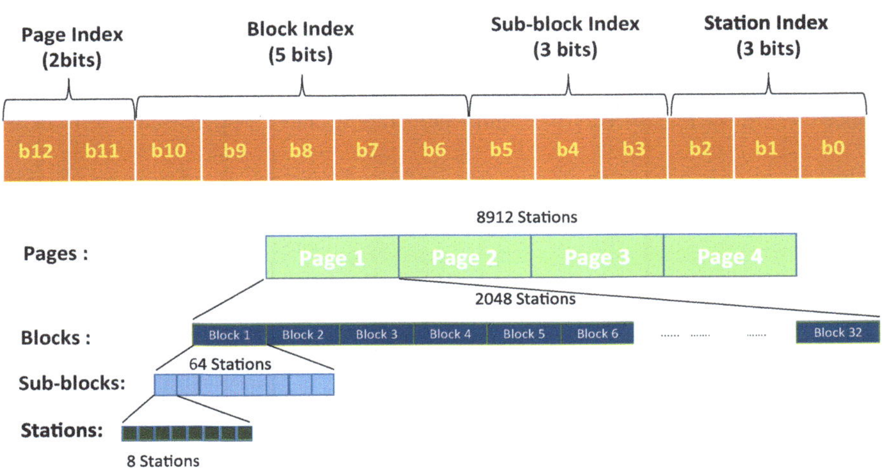

Fig. 3 Hierarchal AID structure

hierarchy. The first two bits are used to organize devices into four pages. The next five bits divide each page into 32 blocks. The following three bits are used to split each block into 8 sub-blocks. The last three bits are to address each device with an index, i.e., each sub-block can have a maximum of eight devices. The devices use assigned AIDs to communicate with the AP [21].

RAW Mechanism

IEEE 802.11ah supports huge number of devices based on the hierarchical AID structure. In a dense network, the probability of collision among the devices increases when all the devices are contending for channel access. Hence, IEEE 802.11ah introduces the RAW mechanism to reduce the collisions and interference among the devices. The structure of RAW is illustrated in Fig. 4. The RAW period is divided into RAW slots, and each RAW slot is allocated to the group of devices. The group of devices contend for the channel access and the remaining devices are prohibited to access the channel [33]. This helps to improve the performance of the dense networks when massive number of devices are contending for channel access. Information regarding the number of RAW slots, slot duration, and assignment of slots are indicated in a separate RAW Parameter Set (RPS) beacon frame which is periodically broadcasted by the AP to all the devices. During the RAW slot, the group of devices accesses the channel using Enhanced Distributed Channel Access/ Distributed Coordination Function (EDCA/DCF) [29]. Each device chooses allocated RAW slot based on the function

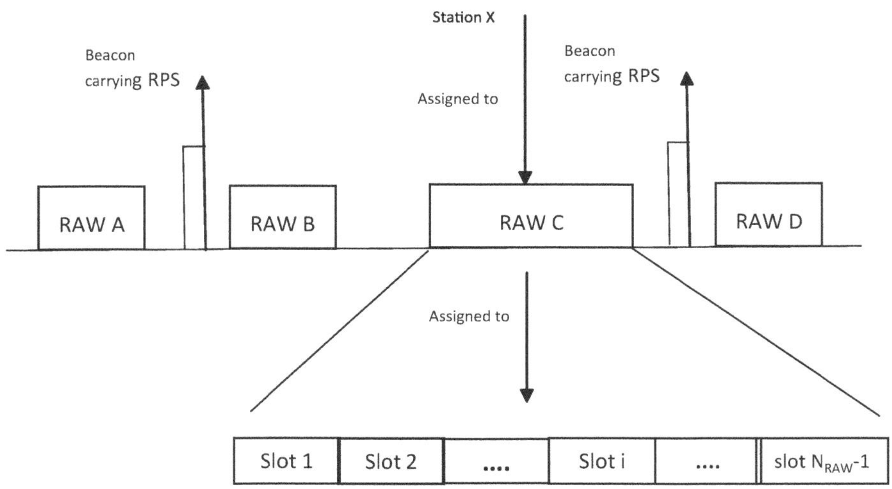

Fig. 4 Structure of RAW

$$i_{slot} = (AID_n + F_{offset})\ mod\ N_{RAW}, \quad (1)$$

where AID_n is the association identifier of the nth device, F_{offset} is to provide the fairness among the devices to choose a RAW slot, and N_{RAW} is the total number of RAW slots.

There are two types of RAW mechanisms: (1) Generic RAW mechanism (2) Triggered RAW mechanism. In the generic RAW mechanism, the AP assigns a RAW slot to every group of devices [22]. But in low-density networks, none of the devices may contend in a RAW slot resulting in unused slots as shown in Fig. 5. Whereas in the high-density network, there will be at least one device ready to contend for the channel. This phenomenon reduces the performance of the RAW mechanism. To overcome this problem, a triggered RAW mechanism is introduced as shown in Fig. 6. The mechanism is activated by enabling the Resource Allocation (RA) frame indication bit in the RAW type field. In this mechanism, the first RAW period is used to contend for slot reservation. The device which deserves a RAW slot is informed using RA frame. In the next RAW period, only the reserved RAW slot appears for contention. For example, the devices that belong to group-1 and group-3 contend for slot reservation in the next RAW period by sending the PS_Poll request, while slot-2 and slot-4 are unused. After the first RAW period, the AP broadcasts RA frame indicating the reservation of RAW slots for the respective devices. Hence only two RAW slots appear in the next RAW period. This mechanism reduces the wastage of channel resources, thus increasing the performance of the RAW mechanism.

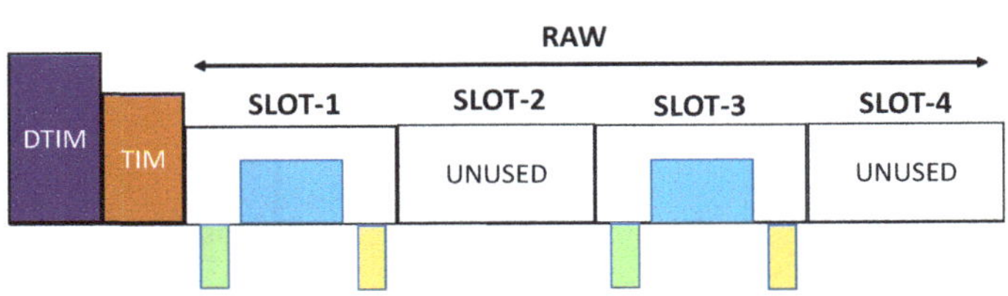

Fig. 5 Generic RAW mechanism

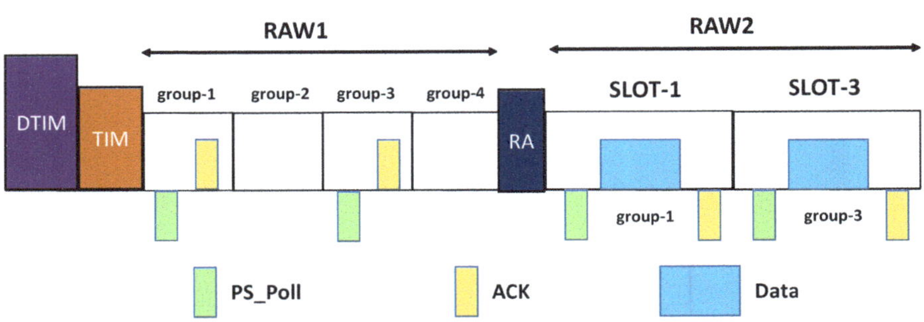

Fig. 6 Triggered RAW mechanism

Apart from this, the RAW mechanism defines two particular cases. First, No Crossing (NCR) slot, i.e., all the transmissions should end before the boundary of the RAW slot. However, this case provides a fair allocation of the channel but also wastes the channel resources [18]. Second, Crossing (CR) slot, i.e., the transmissions are allowed to cross the boundary of the RAW slot which violates the fairness but effectively utilizes the channel resources.

Extended Coverage

IEEE 802.11ah provides the single-hop communication for the range of 1000 m because it operates at widely available sub-1 GHz ISM bands in different countries [31]. As it operates at lower frequency bands, the feature of extended coverage can be supported by IEEE 802.11ah.

Overhead Minimization

Although the IEEE 802.11ah PHY layer provides high data rates, the aggregate throughput of the MAC layer is very low due to the huge overhead of the control packets. To reduce the overhead, IEEE 802.11ah includes several short MAC headers, Null Data Packet (NDP) MAC frames, and short beacons to improve the network performance. Short MAC headers include required fields for transmission. The frames such as Clear To Send (CTS), Acknowledgment (ACK) do not contain any payload information hence to reduce the overhead, IEEE 802.11ah uses NDP CTS and NDP ACK, etc. Several beacon frames have been continuously sent by the AP to all the devices within the coverage area which creates huge overhead. Hence, to minimize the overhead, IEEE 802.11ah introduces short beacons. These are periodically sent by the AP [18].

Power Management

To utilize the power/energy resources efficiently, IEEE 802.11ah uses TIM and TWT frames. The format of AID structure enables the TIM segments to sleep the devices for a certain time when they are in a doze state. This helps in saving the power/energy resources of devices in critical conditions [19].

Rate Adaptation Based on Various MCS

Rate adaptation is a mechanism to choose the data rate based on channel conditions. IEEE 802.11ah supports various MCSs based on the data rates ranging from 0.15 Mbps to 78 Mbps. Hence the devices can choose the data rate based on the channel conditions. This improves the aggregate throughput of the network [18].

Group-Based Mechanism

Grouping is a mechanism to cluster the devices based on the network utilities. Group-Synchronized DCF (GS-DCF) is a group-based mechanism where the group of devices contend for channel access and reduces the contention among devices. As grouping is an inevitable part of GS-DCF, this helps in improving the throughput, fairness, and eliminates the problem of hidden terminals. The IEEE 802.11ah standard does not provide any grouping scheme hence it opens the doors to develop the protocols for grouping of devices in 802.11ah, still it is an active research area [23, 40].

4 Design Issues and Challenges

The main aim of IoV is to integrate various heterogeneous vehicles, things, and networks to provide the ubiquitous connected environment that has to be manageable. So far, we have discussed the features of IEEE 802.11ah and its importance in IoV networks. Due to the versatile requirements of IoV to support various applications, the following issues and challenges are to be considered while designing the IoV based IoT enabled network.

4.1 Management of Hierarchical AID Structure

IoV applications support large number of vehicles in dense networks. These applications operate over a large transmission range in an area. The legacy 802.11 standard can handle 2007 devices per central coordinator due to the restricted number of AIDs [21]. To extend the scalability, IEEE 802.11ah uses a novel hierarchal AID structure as shown in Fig. 3. Hence, it supports 8192 associated devices, thereby maximizing the number of connected vehicles. As vehicles in IoV networks move with very high velocity, assigning a unique AID and management of AIDs in a dynamic scenario is a complex task.

4.2 Assignment of RAW Slots

In dense IoV network scenarios, because of the massive number of vehicles, collisions can take place when all the devices are contending for channel access. It results in the reduction of overall network performance which affects the amount of information shared among the vehicles. Hence, IEEE 802.11ah adopts the RAW mechanism. The concept of RAW mechanism is intended to mitigate the collisions by classifying the devices into various groups and protecting channel access at a particular period for the intended group of devices. Thus, the network performance can be improved [18]. In the IoV network, this can help in reducing channel contention among vehicles by making them into several groups. As a consequence, severe problems such as road accidents can be reduced. But vehicles in IoV scenario have very high mobility hence, assigning each vehicle with a corresponding RAW slot is a challenging issue.

4.3 Formation of Groups

The RAW mechanism allows the devices to form into several groups. Hence, grouping strategy is the inevitable process in the RAW mechanism. The method of grouping reduces channel contention, hidden and exposed terminal problems [11, 37]. Unfortunately, the IEEE 802.11ah task group does not provide any grouping mechanism. It provides the opportunity to bring different grouping strategies as an active research area. In the IoV network scenario, the vehicles will move at varying speeds. Thus, forming the vehicles into groups is a critical task.

4.4 Latency in Association Process

The MAC layer feature of IEEE 802.11ah, namely hierarchical AID structure and RAW mechanism, helps the IoV network to support large number of devices per central coordinator. However, each device has to wait a random amount of time to associate with the central coordinator to share the information among the vehicles. This makes the association process delayed for each vehicle and causes latency in the IoV scenario. Hence, designing a mechanism to reduce the amount of delay in the association process is a challenging issue.

4.5 Performance Anomaly

In a real-time scenario, the vehicles choose different data rates based on channel conditions. Due to this, an unfairness problem exists among the vehicles known as performance anomaly. In contrast, vehicles with the lowest data rate slow the remaining vehicles down to its rate which affects the aggregate throughput of the network. Similarly, the unfairness problem among the vehicles arises due to varying speeds of the vehicles. The fast-moving vehicles will have less resident time to communicate with the central coordinator compared to the slow-moving vehicles. Hence, the mitigation of performance anomaly and unfairness in the IoV scenario is a critical task [14].

4.6 Privacy and Security Issues

IoV integrates several technologies, services, and standards. The increase in number of devices and heterogeneity leads to the vulnerability of data being transmitted among the vehicles. The data should be safe from hackers, malicious and cyber-attacks. Otherwise, they may access the data and can directly control the vehicles which may lead to road accidents. Hence, the privacy and security are needed in every stage of IoV like the privacy of person, data, location, etc. Because of heterogeneous characteristics of IoV, ensuring privacy and security is a challenging task [28].

Acronym	Abbreviation
IoT	Internet of Things
M2M	Machine to Machine
WPAN	Wireless Personal Area Network
LPWAN	Low Power Wide Area Networks
LoRa	Long Range Wide Area
WLAN	Wireless Local Area Networks
ISM	International Scientific Medicine
IEEE	Institute of Electrical and Electronics Engineers
Wi-Fi	Wireless-Fidelity
MCS	Modulation and Coding Schemes
PHY	Physical
MAC	Medium Access Control
AID	Association Identifier
RAW	Restricted Access Window
TIM	Traffic Indication Map
TWT	Target Wakeup Time
ITS	Intelligent Transportation Systems
IoV	Internet of Vehicles
GPS	Global Positioning System
WAVE	Wireless Access in Vehicular Environments
LAN/MAN	Local Area Network/Metropolitan Area Network
GSM	Global System for Mobile Communication
LTE	Long Term Evaluation
QoS	Quality of Service
OFDM	Orthogonal Frequency Division Multiplexing
BPSK	Binary Phase Shift Keying
QPSK	Quadrature Phase Shift Keying
QAM	Quadrature Amplitude Modulation
MIMO	Multiple Input Multiple Output
AP	Access Point
RPS	RAW Parameter Set

EDCA/DCF	Enhanced Distributed Channel Access/Distributed Coordination Function
RA	Resource Allocation
PS	Power Save
NCR	No Crossing
CR	Crossing
NDP	Null Data Packet
CTS	Clear To Send
ACK	Acknowledgment
NDP CTS	Null Data Packet Clear To Send
NDP ACK	Null Data Packet Acknowledgment
GS-DCF	Group-Synchronized DCF

5 Conclusion

In this chapter, we have presented the salient technological advancements and its benefits of IoV with IoT enabled networks. Due to the rapid increment in number of vehicles, the transportation task becomes a complex issue. Consequently, it leads to traffic congestion, road accidents, increased pollution, etc. IoV provides various benefits in ITS that include safe driving, efficient transportation, cloud storage, etc. In a dense network scenario, among the various IoT communication technologies, IEEE 802.11ah is the perfect candidate to reduce the contention among vehicles to access the channel and to support large coverage area. The IEEE 802 task group has proposed 802.11ah with many features such as hierarchal AID structure, RAW mechanism, and various MCSs, TIM, and TWT. However, apart from its advantages, based on the characteristics of vehicles, we have explained the various design and challenging issues of IEEE 802.11ah in the IoV scenario.

References

1. Akpakwu, G. A., Silva, B. J., Hancke, G. P., & Abu-Mahfouz, A. M. (2018). A survey on 5g networks for the internet of things: Communication technologies and challenges. *IEEE Access, 6*, 3619–3647.
2. Al-Fuqaha, A., Guizani, M., Mohammadi, M., Aledhari, M., & Ayyash, M. (2015, fourth quarter). Internet of things: A survey on enabling technologies, protocols, and applications. *IEEE Communications Surveys Tutorials, 17*(4), 2347–2376.
3. Al-Sultan, S., Al-Doori, M. M., Al-Bayatti, A. H., & Zedan, H. (2014). A comprehensive survey on vehicular ad hoc network. *Journal of Network and Computer Applications, 37*, 380–392. http://www.sciencedirect.com/science/article/pii/S108480451300074X
4. Aust, S., Prasad, R. V., & Niemegeers, I. G. M. M. (2015, third quarter). Outdoor long-range WLANs: A lesson for IEEE 802.11ah. *IEEE Communications Surveys & Tutorials, 17*(3), 1761–1775.

5. Baños-Gonzalez, V., Afaqui, M. S., López-Aguilera, E., & Villegas, E. G. (2016). IEEE 802.11ah: A technology to face the IoT challenge. *Sensors, 16*(11), 1960. https://doi.org/10.3390/s16111960
6. Bianchi, G. (2000). Performance analysis of the IEEE 802.11 distributed coordination function. *IEEE Journal on Selected Areas in Communications, 18*(3), 535–547.
7. Chen, M., Tian, Y., Fortino, G., Zhang, J., & Humar, I. (2018). Cognitive internet of vehicles. *Computer Communications, 120*(C), 58–70. https://doi.org/10.1016/j.comcom.2018.02.006
8. Contreras-Castillo, J., Zeadally, S., & Guerrero Ibañez, J. A. (2016). Solving vehicular ad hoc network challenges with big data solutions. *IET Networks, 5*(4), 81–84.
9. Contreras-Castillo, J., Zeadally, S., & Guerrero-Ibañez, J. A. (2018). Internet of vehicles: Architecture, protocols, and security. *IEEE Internet of Things Journal, 5*(5), 3701–3709.
10. Gazis, V. (2017, first quarter). A survey of standards for machine-to-machine and the internet of things. *IEEE Communications Surveys & Tutorials, 19*(1), 482–511.
11. Ghasemiahmadi, M., Li, Y., & Cai, L. (2017, March). RSS-based grouping strategy for avoiding hidden terminals with GS-DCF MAC protocol. In *2017 IEEE Wireless Communications and Networking Conference (WCNC)* (pp. 1–6)
12. Golestan, K., Soua, R., Karray, F., & Kamel, M. S. (2016). Situation awareness within the context of connected cars. *Information Fusion, 29*(C), 68–83. https://doi.org/10.1016/j.inffus.2015.08.001
13. Gupta, M., & Sandhu, R. (2018). Authorization framework for secure cloud assisted connected cars and vehicular internet of things. In *Proceedings of the 23rd ACM on Symposium on Access Control Models and Technologies, SACMAT '18* (pp. 193–204). New York, NY: ACM. http://doi.acm.org/10.1145/3205977.3205994
14. Harigovindan, V. P., Babu, A. V., & Jacob, L. (2014). Proportional fair resource allocation in vehicle-to-infrastructure networks for drive-thru internet applications. *Computer Communications, 40*, 33–50. http://dx.doi.org/10.1016/j.comcom.2013.12.001
15. IEEE Std 802.11ah-2016 (Amendment to IEEE Std 802.11-2016, as amended by IEEE Std 802.11ai-2016). (2017, May). IEEE standard for information technology – Telecommunications and information exchange between systems – Local and metropolitan area networks–specific requirements – Part 11: Wireless LAN medium access control (MAC) and physical layer (PHY) specifications amendment 2: Sub 1 GHz license exempt operation (pp. 1–594).
16. Joy, J., & Gerla, M. (2017, July). Internet of vehicles and autonomous connected car – privacy and security issues. In *2017 26th International Conference on Computer Communication and Networks (ICCCN)* (pp. 1–9).
17. Kang, J., Yu, R., Huang, X., & Zhang, Y. (2018). Privacy-preserved pseudonym scheme for fog computing supported internet of vehicles. *IEEE Transactions on Intelligent Transportation Systems, 19*(8), 2627–2637.
18. Khorov, E., Lyakhov, A., Krotov, A., & Guschin, A. (2015). A survey on IEEE 802.11ah: An enabling networking technology for smart cities. *Computer Communications, 58*, 53–69. Special issue on Networking and Communications for Smart Cities. http://www.sciencedirect.com/science/article/pii/S0140366414002989
19. Kim, T., & Chang, J. M. (2017). Enhanced power saving mechanism for large-scale 802.11ah wireless sensor networks. *IEEE Transactions on Green Communications and Networking, 1*(4), 516–527.
20. Lai, L., & Suda, N. (2018). Enabling deep learning at the IoT edge. In *Proceedings of the International Conference on Computer-Aided Design, ICCAD '18* (pp. 135:1–135:6). New York, NY: ACM. http://doi.acm.org/10.1145/3240765.3243473
21. Lei, X., & Rhee, S. H. (2017). Performance improvement of Sub 1 GHz WLANs for future IoT environments. *Wireless Personal Communications, 93*(4), 933–947. https://doi.org/10.1007/s11277-017-3947-3
22. Mahesh, M., & Harigovindan, V. P. (2018). Throughput and energy efficiency analysis of the IEEE 802.11ah restricted access window mechanism. In P. Bhattacharyya, H.G. Sastry, V. Marriboyina, & R. Sharma (Eds.), *Smart and innovative trends in next generation computing technologies* (pp. 227–237). Singapore: Springer Singapore.

23. Mahesh, M., & Harigovindan, V. P. (2019). Restricted access window-based novel service differentiation scheme for group-synchronized DCF. *IEEE Communications Letters, 23*(5), 900–903.
24. Miorandi, D., Sicari, S., Pellegrini, F. D., & Chlamtac, I. (2012). Internet of things: Vision, applications and research challenges. *Ad Hoc Networks, 10*(7), 1497–1516. http://www.sciencedirect.com/science/article/pii/S1570870512000674
25. Park, C. W., Hwang, D., & Lee, T. J. (2014). Enhancement of IEEE 802.11ah MAC for M2M communications. *IEEE Communications Letters, 18*(7), 1151–1154.
26. Priyan, M. K., & Devi, G. U. (2019). A survey on internet of vehicles: Applications, technologies, challenges and opportunities. *International Journal of Advanced Intelligence Paradigms, 12*(1–2), 98–119. http://dl.acm.org/citation.cfm?id=3302557.3302567
27. S, M. M., & Rao, S. N. (2017, December). A survey of the state of the art of 802.11ah. In *2017 IEEE International Conference on Computational Intelligence and Computing Research (ICCIC)* (pp. 1–4). (2017)
28. Samara, G., & Al-Raba'nah, Y. (2017). Security issues in vehicular ad hoc networks (VANET): A survey. CoRR abs/1712.04263. http://arxiv.org/abs/1712.04263
29. Sangeetha, U., & Babu, A. (2019). Performance analysis of IEEE 802.11ah wireless local area network under the restricted access window-based mechanism. *International Journal of Communication Systems, 32*(4), e3888. https://onlinelibrary.wiley.com/doi/abs/10.1002/dac.3888
30. Siddiqa, A., Shah, M. A., Khattak, H. A., Akhunzada, A., Ali, I., Razak, Z. B., et al. (2018). Social internet of vehicles: Complexity, adaptivity, issues and beyond. *IEEE Access, 6*, 62089–62106.
31. Sljivo, A., Kerkhove, D., Tian, L., Famaey, J., Munteanu, A., Moerman, I., et al. (2018). Performance evaluation of IEEE 802.11ah networks with high-throughput bidirectional traffic. *Sensors, 18*(2), 325. https://doi.org/10.3390/s18020325
32. Sun, W., Choi, M., & Choi, S. (2013). IEEE 802.11ah: A long range 802.11 WLAN at Sub 1 GHz. *Journal of ICT Standardization, 1*, 1–26. http://riverpublishers.com/journal_read_html_article.php?j=JICTS/1/1/5
33. Tian, L., Famaey, J., & Latré, S. (2016, June). Evaluation of the IEEE 802.11ah restricted access window mechanism for dense IoT networks. In *2016 IEEE 17th International Symposium on a World of Wireless, Mobile and Multimedia Networks (WoWMoM)* (pp. 1–9).
34. Tian, L., Khorov, E. M., Latré, S., & Famaey, J. (2017). Real-time station grouping under dynamic traffic for IEEE 802.11ah. *Sensors, 17*(7), 1559. https://doi.org/10.3390/s17071559
35. Wu, W., Yang, Z., & Li, K. (2016). Chapter 16 – internet of vehicles and applications. In R. Buyya & A. V. Dastjerdi (Eds.), *Internet of Things* (pp. 299–317). Burlington, VT: Morgan Kaufmann. http://www.sciencedirect.com/science/article/pii/B9780128053959000162
36. Xu, H., Yu, W., Griffith, D., & Golmie, N. (2018). A survey on industrial internet of things: A cyber-physical systems perspective. *IEEE Access, 6*, 78238–78259.
37. Yoon, S. G., Seo, J. O., & Bahk, S. (2016). Regrouping algorithm to alleviate the hidden node problem in 802.11ah networks. *Computer Networks, 105*, 22–32. http://www.sciencedirect.com/science/article/pii/S1389128616301487
38. Zanella, A., Bui, N., Castellani, A., Vangelista, L., & Zorzi, M. (2014). Internet of things for smart cities. *IEEE Internet of Things Journal, 1*(1), 22–32.
39. Zhang, H., Zhang, Q., & Du, X. (2015). Toward vehicle-assisted cloud computing for smartphones. *IEEE Transactions on Vehicular Technology, 64*(12), 5610–5618.
40. Zheng, L., Ni, M., Cai, L., Pan, J., Ghosh, C., & Doppler, K. (2014). Performance analysis of group-synchronized DCF for dense IEEE 802.11 networks. *IEEE Transactions on Wireless Communications, 13*(11), 6180–6192.
41. Zheng, S., & Lei, Z. (2014, November). TIM encoding for IEEE 802.11ah based WLAN. In *2014 IEEE International Conference on Communication Systems* (pp. 559–563).

Medium Access Control in Connected Vehicles: Advances and Limitations

Raghavendra Pal, Nishu Gupta, Arun Prakash, and Rajeev Tripathi

1 Introduction

The medium access control (MAC) protocols for vehicular communications are different from the MAC protocols used in other applications. These are designed to cater the need of more than one type of applications which are safety related and non-safety related. In addition to it, limited bandwidth requires the protocol to be equipped with cognitive capability, i.e. to switch to other channels of nearby bands. This chapter provides a comprehensive classification of MAC protocols used for vehicular communication. A simple vehicular communication scenario is shown in Fig. 1. Vehicles as well as roadside units (RSUs) are shown in the figure. RSUs are located at the roadsides and help vehicles to communicate with each other but not all the protocols use RSUs. Some major challenges in the protocol implementations in vehicular communications are shown in Fig. 2.

There are several factors which require a separate set of MAC protocols for vehicular communication. First of which is the frequent disconnections due to the rapid movement of vehicles. To mitigate the effect of the rapid movement of vehicles, clustering based MAC protocols were developed. A cluster is a group of vehicles having some almost similar properties such as velocity, position, vehicle type, etc. Each vehicle of a cluster uses the same channel at the same time. The second factor is the presence of 2 types of channels in the band allocated for vehicular communication that is dedicated short range communication (DSRC)

R. Pal (✉) · N. Gupta
Vaagdevi College of Engineering, Warangal, India
e-mail: raghavendrapal@ieee.org

A. Prakash · R. Tripathi
Department of Electronics and Communication Engineering, Motilal Nehru National Institute of Technology Allahabad, Prayagraj, Uttar Pradesh, India

© Springer Nature Switzerland AG 2021

N. Gupta et al. (eds.), *Internet of Vehicles and its Applications in Autonomous Driving*, Unmanned System Technologies, https://doi.org/10.1007/978-3-030-46335-9_5

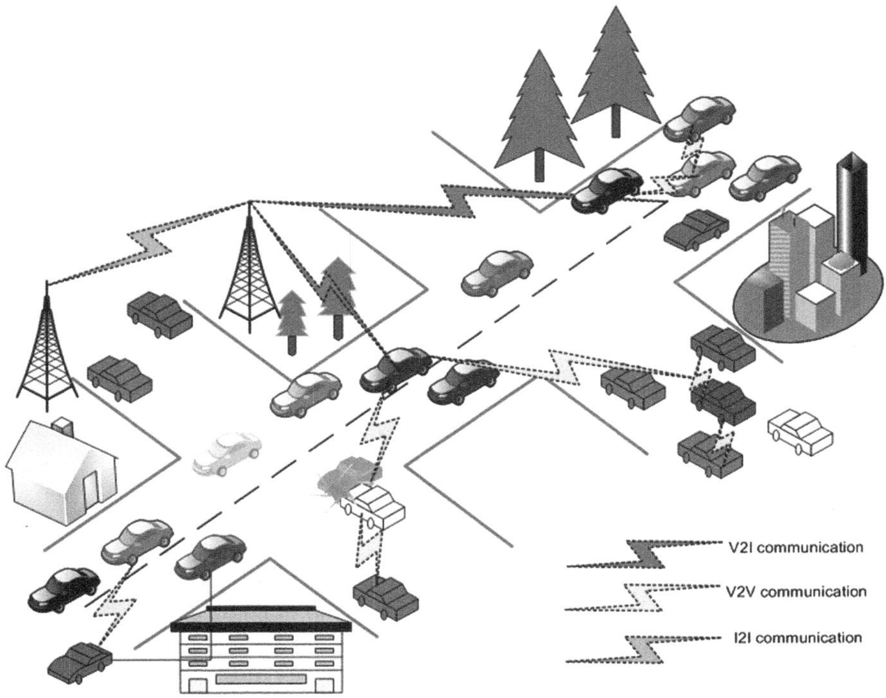

Fig. 1 Vehicular communication scenario

Fig. 2 Challenges in vehicular communications

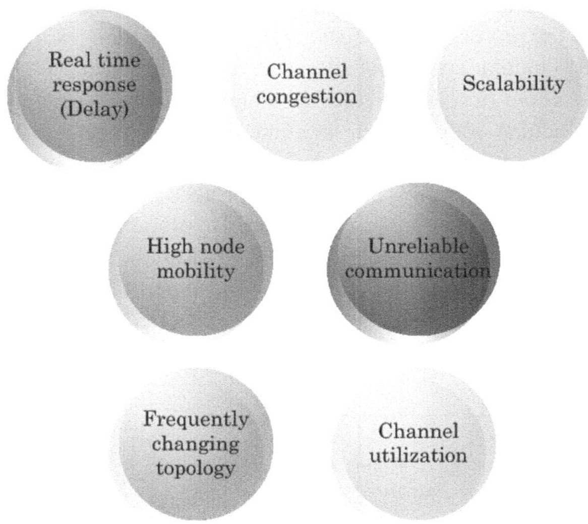

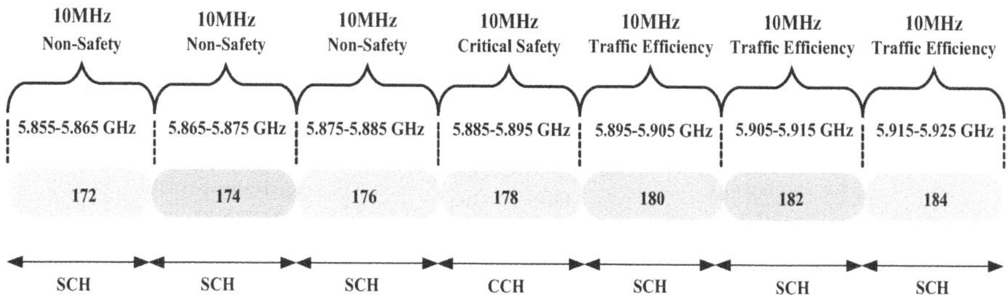

Fig. 3 Dedicated short range communication band

band [1] as shown in Fig. 3. These two types of channels are control channel (CCH) and service channels (SCHs). There are only one CCH and six SCHs.

The time in the network is divided into intervals of 100 ms called synchronization interval (SI). The SI is further divided into 50 ms intervals called control channel interval (CCHI) and service channel interval (SCHI). CCHI is used to transmit safety messages as well as control information using CCH. Infotainment and other non-safety messages are transmitted in SCHI using one of the SCHs. These types of MAC protocols are called multichannel MAC protocols. The nomenclature "multichannel" defines that these protocols provide a provision for vehicles to operate on more than one type of channels. It allows the fair delivery of both types of data, i.e. safety and non-safety. There are further modifications in this classification called the variable interval MAC protocols. In these, the CCHI and SCHI are varied according to local conditions such as number of vehicles and packets in the transmission queues of vehicles [2]. A new type of protocol called triggered protocol emerged recently that enables the immediate switching from SCHI to CCHI whenever a safety packet arrives [3].

The packets in the vehicular communication are classified into safety and non-safety packets. These two classifications are further divided into various classes [4]. Safety packets are life-critical messages which are required to be transmitted immediately. Thus there needs to be some mechanism which helps in defining the priority of messages to be transmitted. The protocols which schedule the messages in the queue according to the priority are called the scheduling based MAC protocols.

Since there are only seven channels allocated for the vehicular communications, these channels can get congested in the dense urban scenarios. In this case nearby bands can also be used for the vehicular communications. Cognitive radio technology [5] helps in this scenario. It enables the vehicles to sense other bands and use those when the primary users of those bands are not communicating. Thus the related protocols are called cognitive radio based MAC protocols.

There are many other types of MAC protocols also. Hence, to understand those clearly, some classification is needed to be done. There are slight variations in different protocols of the same class itself. Major classification of MAC protocols

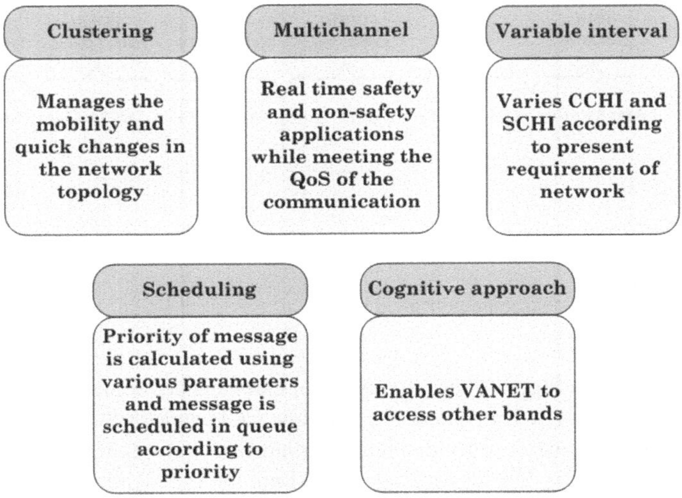

Fig. 4 Major classifications of MAC protocols for vehicular communications

is shown in Fig. 4 and the details of each will be discussed in the next section. It will give a deep understanding of the latest advancements done in this area and the existing limitations.

2 Medium Access Control Protocols for Vehicular Communication

2.1 Clustering Based MAC Protocols

These protocols are required for the purpose of mitigating the effect of rapid vehicle mobility in vehicular communications. The clustering is mostly done on the region basis or mobility basis. However, other bases for clustering are also there though not used widely. Figure 5 shows the basic clustered scenario. Vehicles in a cluster are called cluster members. Rule of being a cluster member is that the vehicle will have to tune to the same channel as of the cluster. Hence, all the cluster members are tuned to the same channel at the same time. Cluster head, on the other hand, can use more than one channel because it has to transmit the message not only to its cluster members but also to nearby cluster heads (inter-cluster communication).

Only cluster head can forward the message to other cluster heads. No other cluster member is allowed for multihop transmission. In region based clustering, the vehicles of a particular region are grouped into clusters. The vehicles which go out of the boundary of the region do not remain as cluster member. These protocols are seldom used than the mobility based clustering protocols. In mobility based

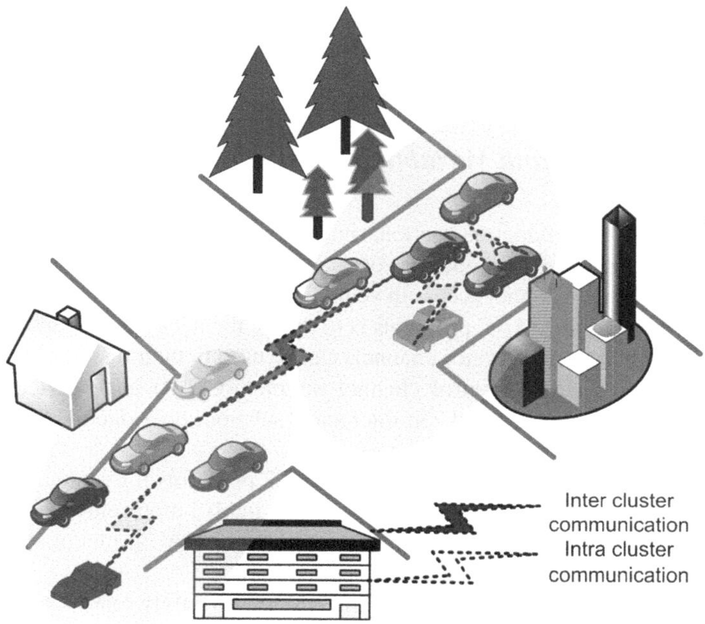

Fig. 5 Clustered vehicular communication scenario

clustering protocols, the vehicle with similar velocity is grouped into a cluster. The vehicle having the least relative mobility in comparison to other cluster members is selected as cluster head. However, the authors in [6] used the relative mobility as well as the distance of the vehicle from center of cluster to select the cluster head. Some protocols use gateways in addition to cluster heads. When a gateway is present, the multihop communication takes place through gateway only and the cluster head is used only for controlling purpose [7]. Some of the protocols use two gateways [8]: one is for forward communication and another is for backward communication. Gateways are mostly at the boundary of the cluster, thus increasing the communication range.

There are several other advantages of the clustering as well. It can reduce the effect of flooding of vehicles since only the cluster head or gateway is allowed for multihop transmission. Another advantage is that the cluster head can control the transmission of vehicles. For these advantages, the clustering is applied with the other types of protocols such as multichannel, variable interval, cognitive, etc.

There are several limitations of clustering based MAC protocols such as vehicles in an underpass and on flyovers can form the cluster. This should not be the case because these vehicles do not affect each other. The second problem is the sudden breaking or acceleration of vehicles. It can cause the breaking of a cluster. Further, increasing number of vehicles on road can result in overcrowding of clusters. This can cause the collision of packets within cluster itself.

2.2 Multichannel and Variable Interval MAC Protocols

As discussed in the introduction section, multichannel MAC protocols were developed since there are more than one type of channels. IEEE 1609.4 [9] is the standard that provides for multichannel access in vehicular communications. The idea behind developing multichannel MAC protocols is to divide the network time into intervals so that control channel and service channels can be used in their respective intervals. These intervals are called control channel interval (CCHI) and service channel interval (SCHI). Figure 6 shows the multichannel schemes provided in IEEE 1609.4. Alternating scheme is the most popular one.

The performance analysis of multichannel MAC protocols for vehicular communication requires a different approach. The authors in [10] represented an analytical model to get the packet delivery ratio of a multichannel MAC protocol in vehicular communication.

However, these schemes are not sufficient enough for safety related applications. Since a message that arrives at the starting of SCHI in alternating method will have to wait for at least 50 ms to be transmitted. Hence, researchers took the initiative of varying the length of CCHI and SCHI according to the neighboring conditions. These conditions are the data available in transmission queues of vehicles, the amount of past safety or non-safety data transmitted, etc. Suppose all the queues of vehicles have a total of 100 safety messages and 50 non-safety messages, then the complete synchronization interval can be divided into 100:50 ratio. It means out of 100 ms, 66 ms can be allocated to CCHI and 34 ms for SCHI. This is just an

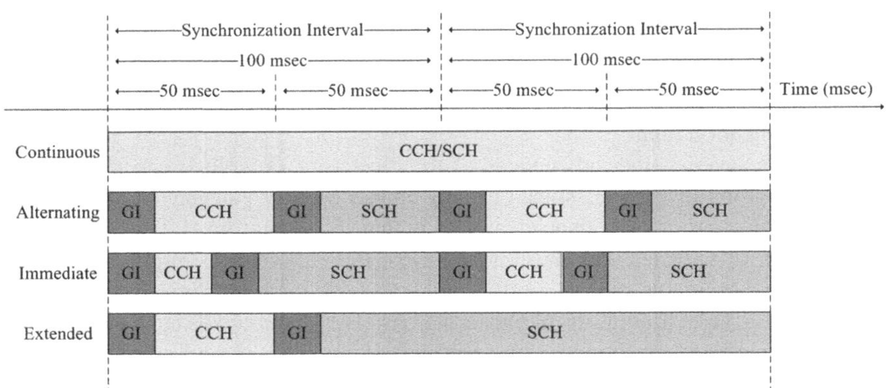

Fig. 6 IEEE 1609.4 Multichannel schemes

example. There are much complex equations used for the calculation of the length of CCHI and SCHI.

These variable interval MAC protocols solve the fixed length interval problem to some extent but do not completely solve the problem. Hidden terminal problem still exists and different vehicles can switch to different channels due to lack of neighborhood information.

2.3 Scheduling Based MAC Protocols

In Fig. 7, left column shows the messages arranged in a queue for transmission. The message at the top will be transmitted first. But there are other messages in queue which are more urgent in nature and are required to be transmitted first such as accident message.

The scheduling comes into play in this scenario. The main idea behind the scheduling based MAC protocols is the most important message should be transmitted first and least important message should be transmitted at last. Hence after rescheduling of queue the messages are as shown in the right column of Fig. 7.

The accident message has come at the top and map download at the bottom of the queue. This priority is assigned according to many factors. These factors are divided into two categories [4]: one is static priority and another is dynamic priority. Static priority consists of only the type or class of message. Classes of the messages are elaborated in Table 1. Dynamic priority depends upon many factors such as speed of transmitting vehicle, deadline remaining of the reception of message, length of message, etc.

The problems associated with scheduling based MAC protocols are the thin line of differentiation between different types of messages. Further, frequent

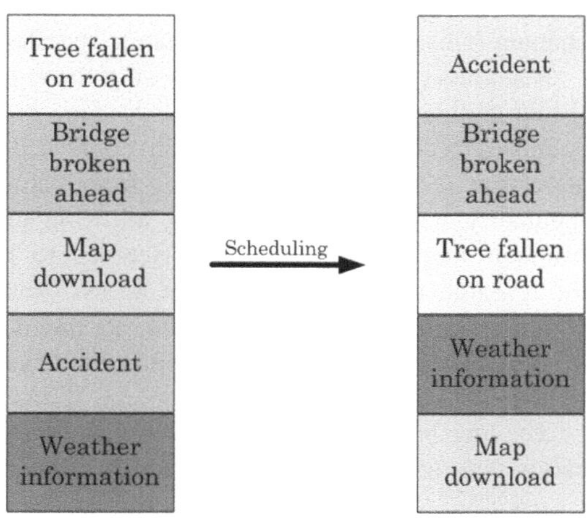

Fig. 7 Transmission queue of a vehicle before and after rescheduling

Table 1 Classifications of vehicular communication applications [11]

Applications	Priority type	Priority class	Network traffic	Message range (m)
Life-critical safety	Safety	Class 1	Event driven	300
Intersection collision avoidance	Safety	Class 1	Event driven	300
Emergency braking avoidance	Safety	Class 1	Event driven	<50
Safety warning	Safety	Class 2	Event driven	300
Cooperative collision avoidance warning	Safety	Class 2	Periodic	300
Electronic toll collection	Non-safety	Class 3	Event driven	300
Automatic parking	Non-safety	Class 3	Event driven	300
Multimedia access	Infotainment	Class 4	Event driven	300
On road service identifier	Infotainment	Class 4	Event driven	0–90

rescheduling of queues can be complex and can cause additional delays. The authors in [12] proposed a hybrid data structure to solve this issue but researchers can provide more efficient data structures.

2.4 Cognitive Radio Based MAC Protocols

Prior to the discussion of cognitive radio based MAC protocols, some introduction to cognitive radio is required. Cognitive radio is a programmable radio that can be dynamically configured to use the best available wireless channel in its neighborhood. It senses and detects a free channel and changes its parameters in such a way that allows the data transmission or data reception on the desired channel. This process is called dynamic spectrum management.

Regulatory bodies found that most of the radio spectrum is inefficiently utilized. While cellular bands are overloaded, some bands are not used efficiently. Some countries performed studies which confirmed this observation. Fixed spectrum allocation restricts the frequencies from being used by other users. Hence the regulatory bodies are considering allowing unlicensed users to access licensed bands if they do not cause interference to licensed users. This has triggered the research in dynamic spectrum access. Software defined radios are the main components in cognitive radio. These are the radios in which hardware components are implemented by software or embedded system. Figure 8 shows the cognitive radio network architecture.

The idea behind applying the cognitive radio technology in vehicular communication is to use the nearby bands for the transmission of non-safety messages and

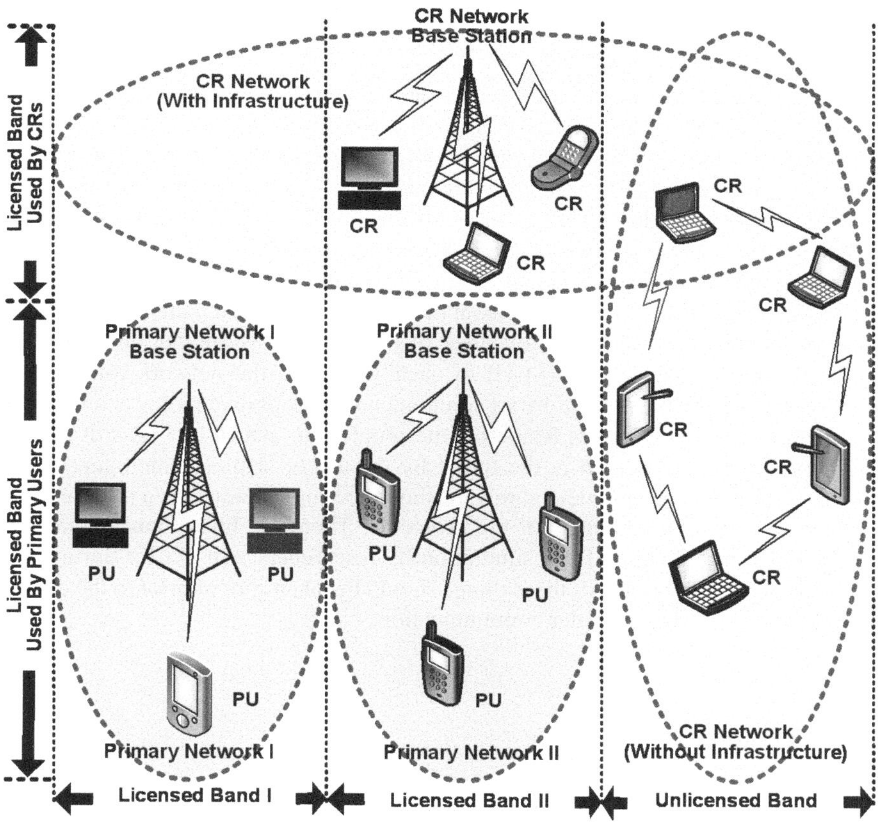

Fig. 8 Cognitive radio network architecture [13]

reserve the whole of DSRC band for the transmission of safety messages [14]. There is much scope in this area since the researchers have recently started working on it.

Vehicles sense the CR channels prior to transmission for their availability and if these are available, vehicles can use these. The major concern in cognitive radio based MAC protocols is the arrival of primary user of a band when the vehicle is using it. It causes a transmission collision and since there is no coordination between PUs and SUs, this problem can only be solved by analyzing the previous history of channel occupancy by PUs. The sensing of CR channels by vehicles can cause additional delays in the transmission. The roadside units (RSUs) can help in this regard. RSUs can be used for sensing the channel proactively and provide these channels to the vehicles which want to transmit the data. In addition to this, RSUs can store the sensed data of the previous few days and process this data and can find the future predictability of CR channels.

Machine learning, artificial intelligence, and deep learning are the areas which can help in the future predictability of CR channels.

3 Discussion

Though significant advances have been made in the area of medium access control in vehicular communications, there are still some research gaps which exist. In multichannel MAC protocols, during SCHI, the selection of a particular SCH out of six available SCHs is a difficult task. The synchronization between all the neighboring vehicles must be made so that all the vehicles select the same SCH. Otherwise the transmission of non-safety messages is not possible. While simulating the network, it seems an easy task since all the vehicles are defined in simulator only. But hardware implementation of it is a really challenging task. In addition to this, synchronization of clock of each vehicle is another challenge. It is required because CCHI and SCHI of each vehicle in the network (or at least the neighboring vehicles) should start simultaneously. Cognitive radios can be used to utilize the nearby spectrum bands but the problem of arrival of PUs still exists. This problem can be tackled in the future by the use of artificial intelligence and machine learning. The complete hardware should be implemented as a pilot project in a city to actual performance of the protocols. There are high chances that the results which are obtained from simulation by researchers will deflect during the hardware implementation. All these things should be taken care of prior to the actual implementation of the vehicular communication.

4 Conclusion

The major classifications of the MAC protocols are presented in this chapter. These are clustering based MAC protocols, multichannel and variable interval MAC protocols, scheduling based MAC protocols, and cognitive radio based MAC protocols. There are some other classifications also but those are kept out of the scope of this chapter. Clustering is done for the mitigation of effect of high mobility and frequent disconnections of vehicles. All the vehicles in a cluster use same channel except cluster head, since it has the capability of multihop transmission. Existing protocols do not take into account three-dimensional scenario and sudden breaking of vehicles can break clusters. Multichannel MAC protocols divide the network time for the use of control channel and service channels in their respective time intervals. The fixed interval of CCHI and SCHI causes problem of resource wastage. For this purpose, variable interval MAC protocols are proposed. These protocols vary the length of CCHI and SCHI according to neighborhood information. Scheduling is done to correctly prioritize the messages to be scheduled in the queue. This improves packet delivery and delay of safety messages. Thin line of differentiation among various safety messages and frequent rescheduling of queues are problems which are needed to be overcome. Cognitive radio in vehicular communication is the area that is not much explored yet. In cognitive radio vehicular communication, low prioritized data are transferred through the channels of other bands that are not used by primary users. Researchers can use the machine learning and related areas for the future prediction of availability of CR channels.

References

1. Fan, B, Stancil, D. D., & Krishnan, H. (2010). Toward understanding characteristics of dedicated short range communications (DSRC) from a perspective of vehicular network engineers. *Proceedings of the Sixteenth Annual International Conference on Mobile Computing and Networking*, ACM, pp. 329–340.https://doi.org/10.1145/1859995.1860033
2. Babu, S., Patra, M., & Murthy, C. S. R. (2015). A novel context-aware variable interval MAC protocol to enhance event-driven message delivery in IEEE 802.11 p/WAVE vehicular networks. *Vehicular Communications, 2*(3), 172–183. https://doi.org/10.1016/j.vehcom.2015.07.001.
3. Pal, R., Prakash, A., & Tripathi, R. (2018). Triggered CCHI multichannel MAC protocol for vehicular ad hoc networks. *Vehicular Communications, 12*, 14–22. https://doi.org/10.1016/j.vehcom.2018.01.007.
4. Dubey, B. B., Chauhan, N., Chand, N., & Awasthi, L. K. (2016). Priority based efficient data scheduling technique for VANETs. *Wireless Networks, 22*(5), 1641–1657.
5. Arslan, H., & Mitola, J., III. (2007). Cognitive radio, software-defined radio, and adaptive wireless systems. *Wireless Communications and Mobile Computing, 7*(9), 1033–1035.
6. Pal, R., Gupta, N., Prakash, A., & Tripathi, R. (2018). Adaptive mobility and range based clustering dependent MAC protocol for vehicular ad-hoc networks. *Wireless Personal Communications, 98*, 1155–1170. https://doi.org/10.1007/s11277-017-4913-9.
7. Ren, M., Khoukhi, L., Labiod, H., Zhang, J., & Vèque, V. (2017). A mobility-based scheme for dynamic clustering in vehicular ad-hoc networks (VANETs). *Vehicular Communications, 9*, 233–241.
8. Chen, Y., Fang, M., Shi, S., Guo, W., & Zheng, X. (2015). Distributed multi-hop clustering algorithm for VANETs based on neighborhood follow. *EURASIP Journal on Wireless Communications and Networking, 2015*, 98.
9. IEEE. (2009). IEEE draft amendments for wireless access in vehicular environments (WAVE). IEEEP802.11p/D5.0.
10. Campolo, C., Molinaro, A., Vinel, A., & Zhang, Y. (2012). Modeling prioritized broadcasting in multichannel vehicular networks. *IEEE Transactions on Vehicular Technology, 61*(2), 687–701. https://doi.org/10.1109/TVT.2011.2181440.
11. Gupta, N., Prakash, A., & Tripathi, R. (2015). Medium access control protocols for safety applications in vehicular ad-hoc network: A classification and comprehensive survey. *Vehicular Communications, 2*(4), 223–237.
12. Pal, R., Prakash, A., Tripathi, R., & Naik, K. (2019). A preemptive priority and hybrid data structure based scheduling mechanism for CR-VANET. *IET Communications., 13*(20), 3443–3451. https://doi.org/10.1049/iet-com.2019.0574.
13. Kumar, K., Prakash, A., & Tripathi, R. (2016). Spectrum handoff in cognitive radio networks: A classification and comprehensive survey. *Journal of Network and Computer Applications, 61*, 161–188.
14. Gupta, N., Prakash, A., & Tripathi, R. (2016). Clustering based cognitive MAC protocol for channel allocation to prioritize safety message dissemination in vehicular ad-hoc network. *Vehicular Communications, 5*, 44–54.

Part III
Advance Driver Assisted System

An Overview of ADAS in Internet of Vehicles

Surjeet and Priyanka Bhardwaj

1 Introduction to Internet of Vehicles

Worldwide network that connects all small objects together is IoT. Though in IoT, all the things are able to communicate with each other. When we restrict these small things connected over internet to vehicles, we name it as Internet of Vehicles (IoV). As the urban population is increasing, ownership of vehicles is also increasing at exponential rate and becoming a big problem. So, there is a great need of IoV based traffic management solution. As the internet is global phenomenon and mostly all devices are internet friendly, so it is quite easy to manage the transportation system with IoV. For sorting out this problem, traditional methods that were employed are speed with CCTV cameras, making use of speed trackers and periodic pollution checks. But when the number of vehicles increases, effectiveness of handling the incoming breaches decreases and hence IoT comes to rescue. There are several methodologies for traffic management using wireless sensor networks. IoT in traffic management provides significant advantages in handling traffic and is currently deployed in construction of smart cities infrastructures.

In IoV based traffic management system to work as in reality, there should be seamless interaction between various vehicles, i.e. they are able to communicate and work together. This can be established with various methodologies. IoV offers huge advantage in traffic management as they provide effective traffic control. IoV provides human proximity detectors, i.e. proximity sensors onboard detect human temperature on road as visibility is quite reduced at night and this is prone to accidents. So, to prevent accidents, these sensors detect human presence and the speed of the vehicle is reduced immediately. Other advantage of IoV

Surjeet (✉) · P. Bhardwaj
Department of Electronics and Communication Engineering, Bharati Vidyapeeth's College of Engineering, New Delhi, India
e-mail: surjeet.balhara@bharatividyapeeth.edu; priyanka.bhardwaj@bharatividyapeeth.edu

© Springer Nature Switzerland AG 2021

N. Gupta et al. (eds.), *Internet of Vehicles and its Applications in Autonomous Driving*, Unmanned System Technologies, https://doi.org/10.1007/978-3-030-46335-9_6

includes theft avoidance, accident avoidance, emergency response, and autonomous vehicles. Concept of IoV requires the installation of processor and sensor on each vehicle. Apart from this, intermediate processors and server should handle the load efficiently, only then the system would be of use. Hence if we enable internet on each and every vehicle on the road, we can establish complete automation of vehicles and traffic. Further we can extend the concept of IoV to all modes of transport and using IoV over traditional IoT is quite advantageous. Thus, Intelligent transportation systems (ITS) provides better architecture and strategies for management of road traffic along with providing effective monitoring and emergency response to traffic incidents.

IoV mainly focuses on ITS and telematics system, these two things can be taken as two main focuses on IoV. ITS is meant for integrated application of communication, information, and control processing technology which is employed in transportation system. On the other hand, telematics refers to the combination of telecommunications and informatics. So, we can consider it as on-board system service platform which provides entertainment, navigation, and various conveniences for automobile users through satellite navigation system and wireless communication technology. Since our automotive field is continuously growing and is having remarkable achievement in improving comfort and safety. Every newly launched vehicle provides ample of new features for improving road safety. Some of the features include Long-Term Evolution (LTE), Wi-Fi hotspots, smart phone connectivity, high resolution touch screen, etc. moreover they have also obstacle detector cameras to avoid collision. Some of the examples may include lane centering sensors in cars like Ford Fusion, Lincoln MKZ, Toyota Prius, etc. Still after such advance features, vehicles are prone to accidents. Reason for the same may be high speed, bad weather (rain, fog), carelessness, etc. The scenario will be improved soon as we are going to have connected vehicles as they are encouraged by governments as well as automotive industries. Along with this, we have got communication between the vehicles with the development of Vehicular Ad hoc Networks (VANETs). For example, in 2015, general motors have developed 4G LTE-connected vehicles and have predictions for fully autonomous vehicles by the end of this decade. LTE connectivity has provided a leading path for concept of internet connected vehicles which is basically an extension of VANETs. VANETs provides with a base for vehicular communications and further provides enhancement in driving experience by improving security, infotainment, and robustness. But in spite of all these measures, VANETs have not gained much industrial interest commercially. Many researchers tried contributing and exploring this concept, but due to many security and privacy related issues, we have not got this to deployment stage. But the considerable amount of research in this field has led to shifting from conventional VANETs to IoV. We have realized this shift through merging VANET with concept of internet of things (IoT), which is currently the most tempting area these days. However, till now, various things have not been decided appropriately such as security framework, potential architecture, and trust models for IoV. In this chapter, starting with the introduction, we will explain the basic concept and components of IoV network. We will also deal with various layered architecture and benefits of IoV networks. We would discuss about the needs of IoV security challenges in IoV networks and approaches to achieve security. Further we will cover trust models and trust establishment approaches for IoV network, many other interesting facts and approaches of IoV with final conclusion.

1.1 Basic Concept of IoV

Internet of vehicles is an emerging field that integrates VANETs, IoT, and mobile computing. In case of IoV, we consider vehicles as smart nodes equipped with internet that enables smart vehicles to collaborate among each other for communication and data sharing. These smart vehicles may also interact with road users like pedestrian and roadside units. They share and gather the information among their surroundings.

IoV offers vast potential and leads to rapid growth of IoV market. Many car manufacturers like BMW, Tesla have already provided with cars having self-driving features. Apart from this various companies like Uber, Google, etc. are also trying to provide with the self-driving mechanism. However, with the rapid increase of IoV, we have got new security challenges. Millions and millions of vehicles are connected via IoT and IoV network uses the same feature and hence provides road safety and avoids malicious attackers, etc. There can be an individual malicious attacker who could attack IoV environment. Apart from this, adversaries may exploit the weakness of IoV like insecure communication channels cause traffic accidents via triggering malicious information. Moreover, attackers can use this medium for launching a targeted attack on military personnel or high rank officials. Various other categories of attacks can also be setup in vehicular system like malware infection affecting the engine and brake of vehicles. Trustworthiness or in other words we can say reliability of data transmitted between the IoV entities.

1.2 Components and Interactive Modes in IoV Network

Internet of vehicle consists of the following components:

- Roadside Units (RSU)
- Vehicles
- Sensors
- Infrastructure
- Personnel devices and humans

Infrastructure and Vehicle Information (I & V)
It allows vehicle to communicate with nearby fire station, hospital, police station, or any other emergency need. Apart from this, vehicles may communicate with home, office, etc. to control the devices such as air conditioner, television, etc.

Sensor and Vehicle Interaction (S & V)
In this, vehicles talk to sensors on signs of traffic light, bus stops, and moreover with the sensors that are embedded in the road for getting the traffic updates and other rerouting alerts. Along with all this, vehicle sensor senses any events in its neighborhood and accordingly provides with alert to the vehicles regarding any collision avoidance, lane departures, etc. Moreover, there are some car engine sensors, for instance, engine speed sensor, massive air flow sensor, fuel temperature sensor, etc.

RSU and Vehicle Information (R & V)
In this, vehicles basically have interaction with the fixed roadside infrastructure which are connected to the internet and provides various facilities that may be related to entertainment, etc.

Vehicle-to-Vehicle Interaction (VTV)
One vehicle may directly interact with other vehicle and exchange the information like regarding brake status, position of vehicle, speed, etc. and accordingly supports with driving alerts within a particular range.

Human and Vehicle Interaction (H & V)
It allows the vehicle to have easy and convenient interaction with the bicyclists and pedestrians on the way and convey the intentions to the neighborhood and accordingly people act in return.

Personnel Devices and Vehicle Interaction (P & V)
In this, we have interaction between vehicles and smart devices like laptop, personnel digital assistants, etc. that is present in device and hence collects the information or may access the media of the user's phone, etc.

Cloud and Vehicle Interaction
All the information passes through a central hub called cloud. Vehicles do have limited computation and storage resources that may not be sufficient for emerging applications, for instance, in vehicle entertainment, etc. Since they require big storage and complex computations, for this vehicle to cloud computation helps by providing proficient support to these applications. In this, vehicle to cloud communication, any node can directly communicate. This call to cloud-based sensors (CBS) include many applications like data security, entertainment, communication, etc. In the interaction, node may access cloud by protocols for instance HTTPS, RPC, HTTP, and direct API calls.

1.3 Architecture of IoV

It is quite a challenging task to have flow design of layered architecture for a worldwide network that includes various other networks. This task requires the identification and grouping of elements of different networks having similar function as a single layer. Many researchers for instance [1–3] provide different approaches for the layered architecture of IoV via various models and approaches. But models with a smaller number of layers were not able to provide detailed elaboration of the actual protocols and functioning of various layers, whereas models with a greater number of layers supported interaction with the maximum technologies along with reducing the complexity of each layer.

Architectural Benefits of IoV

Previously there was no requirement of internet for vehicles. But now the scenario has been changed and most of the models of cars can be paired with smart phones easily via Bluetooth so that user may attend calls, play music, etc. In a similar manner, internet is connected in upcoming cars resulting in their remote controlling. Along with all this, IoV architecture enables vehicles to "X" Communication here "X" could be vehicle, RSU, infrastructure, personnel devices, sensors, human, etc., thus providing benefit to the driver such as increased comfort, better convenience, and increased efficiency and safety.

- *Increased Comfort*

V2I gas allowed the driver to remotely access their home devices such as air conditioners, blowers, etc. before reaching back home.

- *Better Convenience*

V2P has enabled the entertainment system of car to be connected to driver's phone with the help of phone to your media collection at home.

- *Increased Safety and Efficiency*

V2R and V2V communication has increased the safety of the drivers by using features like collision preventions, traffic jams, accidents information given by nearby vehicles or road infrastructure or cloud to save travel time.

1.4 Security Challenges in IoV

IoV have got the property of self-organizing themselves which are connected by internet and have got highly dynamic topology. But this internet connectivity makes the node as threat vulnerable. Thus, the information that the nodes provides is subjected to risk of malicious activity such as sleeting, malicious routing, and eavesdropping. These issues may cause catastrophic consequences such as accidents, etc. and hence security of the complete network is compromised. Apart from security, in IoV networks, safety is a major concern as the precious life of human beings is always at risk. In contrast to this feature, the other previous traditional methods never included safety as their prime concern. Except this, security related issues; there are other unique security related challenges because of many unique features of IoV such as high mobility, short connection time, and frequent disconnections. These properties presents security challenges such as data protection, trust group formation, and certificate management. In order to cater these challenges, we have needed to define security challenges for IoV which is quite challenging. Trustworthiness of traffic information is disseminated by other vehicles or infrastructure. It allows the malicious user to broadcast traffic jam/accident warnings to clear own route and getting time at expense of others. IoV also provides with remote control to the hackers, via these hackers may remotely control the automatic braking system by sending fake V2X message signals or by altering the safety-critical information regarding any vehicle. But anyone now can now misuse the internal networks of vehicles via standard on-board device interface. Apart from this, it can also update the ECV firmware. IoV further provides with automated vehicle identification that further allows the vehicle to recognize itself for accessing to a toll payment or parking. In IoV network, lack of security support the hackers in stealing personnel data such as their financial details.

All the examples support and demonstrate that IoV makes the vehicle openly connected to external networks like internet and further makes it accessible remotely to hackers and other malicious users. Hence, we can say that apart from safety, there is a need of security and privacy to protect them from attacks in which malicious user may take control of the vehicle by stealing data.

Security Schemes for IoV Network

It includes the security approaches for IoV network that has encryption-based security schemes and trust-based security schemes.

Encryption Based Security Schemes

These schemes are further categorized into two categories. First one is symmetric encryption scheme (SES) and the other one is public key encryption scheme (PES). Here SES's show the single key for both purposes, i.e. encryption as well as decryption, whereas on the other hand, the second category, i.e. PES has got different keys, one for encryption and other one for decryption.

SES additionally provides us the benefit of computational complexity. It has got less computational complexity and are fast as compared to PES. It has got delay in the reception of safety related messages which makes them meaningless. So, in order to reduce the delay in reception of safety messages and for achieving more security in IoV networks. This SES has got some problems such as complex algorithm for key exchange SES is quite complex, communication load in these increases due to use of authentication code and confidentiality problem related to key. In case of IoV, nodes are moving freely on roads and thus are prone to security threats. Considering all the issues discussed above, we prefer PESs over SESs as they are more scalable, and they do not include any complicated key management algorithms. Apart from this, node authentication can be conveniently achieved to guarantee the security of the whole network.

Mostly used public key encryption algorithms for IoV include Rivest, Shamir, and Adleman algorithm (RSA) and Elliptic Curve Cryptography algorithm (ECC). Computation complexity of PES is quite high. In order to summarize, we can say both PES as well as SES have their own pros and cons; but none of them can completely solve IoV network security issues. Encryption schemes may be utilized as the hard security solutions that can provide security by non-repudiation, confidentiality, and data integration, etc. Hard security solutions have got binary type situation of 1 or 0, i.e. node either passes the security check or fails. But hard security can detect the regular switching behavior of node for a specific time and then pass the security check and thereafter due to selfish intentions may turn into bad nodes. Thus, we can conclude that hard security schemes are not able to provide reliability and trustworthiness of nodes along with the information sent by them.

Encryption schemes are considered as hard security solutions that provide the security, data integrity, and confidentiality, etc. like binary solutions, in hard security checks, nodes either pass or fail. Hard security checks would not provide reliability and trustworthiness of nodes as well as information sent by them.

2 Introduction to Advanced Driver Assistance Systems

Advanced driver assistance systems, or ADAS, is the term used for development of well-being capacities intended to improve driver, traveler, and walker security by lessening both the seriousness and in general number of mishaps. ADAS can caution drivers of potential risks, intercede to enable the driver to stay in charge so as to forestall a mishap, and, if important, decrease the seriousness of a mishap in the event that it can't be evaded. To put it plainly, ADAS makes up for our missteps, be they heedlessness, incorrect control inputs or, to a limited extent, absolute idiocy.

To state ADAS is turning out to be "progressively normal" is, truth be told, somewhat deceitful. ADAS has been assuming a significant job in the advancement of more secure vehicles for certain decades; however, the early manifestations of car driver-help innovation are presently so all around created, compelling and commonplace that we frequently underestimate them. The most typical and well-known of everything is the non-freezing stopping mechanism, or ABS. Different manifestations of ABS have been created and trialed via car producers since the center of the twentieth century. Once more, the demonstrated advantages of every framework are notable, and ESC, similar to ABS, is presently compulsory in the EU. Indeed, ESC is generally recognized similar to the absolute most significant supporter of vehicle well-being because of its capacity to diminish the number and seriousness of mishaps, explicitly those including a vehicle turning over and leaving the street—a situation that frequently brings about genuine wounds or even fatalities. In 2016 ESC's designer depicted it, maybe with reasonable inclination, as "the absolute most significant piece of ADAS and the reason for every single future framework" (https://spectrum.ieee.org/vehiclesthat-think/nerd life/profiles/van-zattan-wins).

2.1 *Need for ADAS*

To put it plainly, ADAS implies more secure streets. As of now referenced, the job of ADAS is to diminish street passing and wounds by cutting the quantity of street mishaps generally and lessening the seriousness of those that can't be stayed away from. Various examinations have just gone some approach to demonstrating the adequacy of ADAS in true driving circumstances, and the general agreement, both inside the car business and in more extensive authoritative circles, is that the innovation ought to be advanced and embraced on as expansive a scale as could reasonably be expected—and in the near future.

A 2016 report reasoned that a speculative (and as a matter of fact ridiculous) 100% selection of forward impact notice, vulnerable side recognition, and path takeoff cautioning by 2020 would lessen motorway mishaps by 16.3% and mishaps on different streets by 11.6%. Considering progressively complex frameworks, for example, path keeping help, independent crisis braking, night vision and multi-include bundles of consolidated ADAS works, these figures rose to a 45.4% mishap decrease on motorways and a 27.5% decrease on different streets (www.swissre.com/library/The_future_of_motor_insurance.html).

All the more all things considered and considering the anticipated genuine take-up of ADAS bundles that stay, for the time being, generally costly alternatives or standard just on increasingly exorbitant premium models, Swiss Re presumed that a general mishap decrease of 4.3% could be normal by 2020. Different investigations have seen protection guarantee information, which incorporates minor mishaps that bring about harm to vehicles yet not really any mischief to inhabitants, people on foot or other street clients (www.volvocars.com/uk/about/our-developments/city-safety).

It is as of now the case that Euro NCAP's "security help" tests evaluate forward impact notice/self-ruling crisis stopping mechanisms (in different situations), just as path takeoff notice frameworks, path keep help frameworks and speed help frameworks, and it remembers appraisals for these capacities for its general evaluation of new vehicles. Typically, different makers will unavoidably stick to this same pattern by embracing the new innovation. It should not be thought little of how powerful this buyer confronting program has proved to be helpful to purchaser familiarity with accessible new advancements, and ADAS is relied upon to profit comparably with quickened take-up over the coming hardly any years. As referenced before, a few ADAS capacities are as of now compulsory on new vehicles in different districts far and wide, with others set to turn out to be so soon.

Likewise, a rundown of 11 new security highlights was distributed in 2018 that needs commanded on new vehicles by 2021. The rundown incorporates a few existing ADAS capacities, explicitly self-governing crisis braking, path keeping help, insightful speed help framework, and a switching camera/back identification framework. For the entirety of ADAS's advantages, anyway an admonition ought to be made. While ADAS improvement is propelling constantly, it is oneself driving or self-governing autos of things to come that are demonstrating especially capable at standing out as truly newsworthy. This has driven a few fourth of the business to require a clearer separation between cutting edge driver help frameworks and completely self-sufficient autos (http://news.thatcham.org/pressreleases/carmaker-utilization of-the-word-self-sufficient a-threat to-uk-streets 2537576).

It is significant, at that point, that customer attention to the advantages of ADAS increments so as to quicken the take-up of what are for the present, by and large, just discretionary security frameworks. However, there is a conspicuous pathway driving from the present progressing ADAS improvements to the completely self-ruling autos of tomorrow, yet undoubtedly until further notice, the usable word here is "help": in all cases, consistently and regardless of what the degree of help, the driver remains in generally speaking control as well as liable for what the vehicle does.

3 ADAS Future

ADAS innovation is progressing at a practically remarkable rate. It was stated that "the automobile business will change more in the following 5 to 10 years than it has in the previous 50" (www.forbes.com/destinations/joannmuller/2016/01/18/davos-2016-gm-supervisorsees-a-transformation/in-individual versatility/#230e04c746bf). At the core of this announcement is the drive towards making more secure autos, and more secure streets, through ADAS and self-driving innovation.

As an expansion of V2V, V2I will furnish vehicles with data from the street system's framework, for example, traffic lights and signals, variable speed cutoff points, and blockage data. Such data is required to improve security as well as diminish clog by empowering a more liberated progression of traffic, and it is additionally perceived as a key driver towards full self-sufficiency. V2X, in the interim, includes information streams from past the quick street arrange, including cloud-put away data, meteorological updates and perhaps cyclists, walkers and other helpless street clients (VRUs).

Be that as it may, all together for the associated vehicle to be understood, an agreement despite everything should be come to on which correspondence framework the business will receive so as to empower V2V, V2I or V2X. Initially it appeared as though an institutionalized framework called devoted short-run interchanges (DRSC) would win favor. Working in a 75 MHz-wide band around

5.9 GHz, DRSC would permit V2V correspondence and a V2I interface by means of committed side of the road reference points (in spite of the fact that it is muddled who might pay for, work, and look after them).

Meanwhile, V2V's rollout has been obfuscated by the appearance of an elective framework with more noteworthy potential advantages. C-V2X ("C" meaning "cell") was imagined utilizing the LTE 4G cell phone arrange. LTE not just permits direct gadget to-gadget correspondence (for example, not by means of the cell phone system's correspondence foundation), however, it can likewise, critically for V2V and V2I, communicate from one gadget to various beneficiaries. C-V2X additionally opens up the chance of fusing walkers and cyclist information in the V2X circle through cell phone combination, along these lines upgrading VRU security. It ought to be noticed that as far as both equipment and foundation, C-V2X's LTE innovation has no similarity with DRSC—the two frameworks are, as a result, fundamentally unrelated. Associated ADAS without a doubt opens up another universe of chance in improving street well-being and makes ready for full self-rule, however, until an innovation pathway is settled upon, it is across the board reception stays some way off.

4 ADAS Hardware

Current ADAS consolidates various innovations so as to work. Basically, every framework involves in any event one sensor to screen given parameters and hand-off when essential data which is then handled and broke down previously, whenever required, sending an order, for instance. With certain frameworks, a mix of a few or the entirety of the above may happen. Different cautions are essentially progressively obvious: a stopping sensor's inexorably critical rings are today commonplace to numerous drivers, while a forward impact notice framework regularly joins raising admonition tolls with an enlightened dashboard alert and a haptic alarm through the directing wheel.

Haptic cautions—vibrations through the guiding haggle/seat—are likewise a typical component of path takeoff notice frameworks. At the point when the path takeoff cautioning capacity is extended to path keeping help, the framework can intercede in the control of the vehicle to move the vehicle back towards the focal point of its planned path. The car business' ongoing discount change to electric force helped controlling (EPAS), which was embraced generally for eco-friendliness reasons, has furnished numerous vehicles with equipment that is unmistakably appropriate for ADAS works that mediate with self-sufficient directing data sources. Self-leaving capacities and the most recent automobile overload help includes that can control the vehicle for constrained periods in moderate moving rush hour gridlock have likewise been empowered by the appearance of EPAS.

Self-governing braking intercessions are encouraged by segments that proclaimed the appearance of ADAS in any case: the ABS control module and siphon. The present steadiness control, independent crisis braking, versatile voyage control, and congested driving conditions help capacities, among others, utilize advancements of ABS equipment that take into consideration accurately controlled braking intercessions when required. In the event that vital, motor yield is controlled or adjusted by at least one method. On account of footing control and electronic steadiness control frameworks, which regularly require a generally unpretentious constraining of torque, it very well may be finished by stifling the start flash or confining the fuel stream by means of the fuel infusion framework.

4.1 ADAS Sensors

The sensor is required to recognize everything the driver can see as well as that which the driver cannot—or has not took note. There are various types of sensor as of now being used, each with their own points of interest and weaknesses as far as ability, cost, and bundling, and it is progressively the situation that more than one sort of sensor is utilized for everyday work. Each kind of sensor has recognized qualities and shortcomings, so by joining various advances it is conceivable to refine the ADAS capacities. This combination of sensor innovation is quickly turning into the standard—the assignment at that point is to process the convergence of information from various sources both precisely and rapidly.

The developing take-up of ADAS and the progressing improvement of self-ruling vehicles is driving the development of sensor innovation at a quickened rate. As far as item discovery and order, numerous frameworks as of now being used are as yet working at a generally fundamental level and there is far to go before ADAS usefulness can make the hop to completely self-governing applications. Current frameworks can, for instance, battle to recognize people on foot past a quite certain structure. They may neglect to perceive an individual wearing attire that fundamentally adjusts their framework, in the event that they are conveying an enormous article or in the event that they are beneath a specific tallness. As the innovation grows, be that as it may, these constraints will definitely be tended to. Current ADAS sensor innovation can be separated into four primary classifications, which we will take a gander at in more fine-grained detail.

1. Radar

   Maybe the best-perceived of the considerable number of innovations at present received for ADAS sensors is radar. Radar has been used in car frameworks for certain years now, so the equipment is very much evolved and moderately reasonable, making it appealing to vehicle makers.
2. Ultrasonic

   Ultrasonic sensors have a generally short viable working reach—around 2 m—so they are ordinarily utilized in low-speed frameworks. Their utilization in stopping sensors has been far reaching for quite a while, yet they have likewise discovered a spot in progressively complex ADAS capacities, for example, park help, self-stopping, and some vulnerable side observing applications. Ultrasonic sensors are financially savvy and moderately hearty and dependable, in addition to they are unaffected by evening or other testing light conditions, for example, brilliant, low daylight.

3. Lidar

   The word Lidar is a combination of light and radar. It stands for light detection and ranging. There are two essential kinds of Lidar; strong state Lidar, of which a few varieties are being created. One flames a solitary laser through an optical staged exhibit so as to coordinate the shaft in different ways, while the other, alleged glimmer Lidar, utilizes a solitary heartbeat, or blaze, of laser to make its picture. Truly Lidar has been restrictively costly for use underway car applications, however, it is gradually getting increasingly regular in ADAS improvement as the innovation is refined and costs descend. Model completely self-sufficient vehicles have just utilized the cumbersome rooftop mounted Lidar frameworks to great impact.

4. Cameras

   Camera-based arrangements have picked up footing as the ADAS engineer's sensor innovation of decision. They have their constraints—to be specific their powerlessness to traded off execution in poor climate and low or testing light conditions—yet the innovation, while generally new contrasted and, state, radar or ultrasonic sensors, is as of now proficient and adaptable. Not at all like different sensors here, cameras are the main ones ready to recognize shading and difference data, which makes them undeniably fit to catching street sign and street checking data, and they likewise offer the goals to group items, for example, people on foot, cyclists, and motorcyclists. Cameras are additionally incredibly practical, which makes them especially alluring to volume-selling vehicle producers. Because of the restrictions of the innovation, the information from cameras sensors is progressively being joined with radar to give an increasingly hearty and solid information stream over a more extensive assortment of conditions.

   Cameras are utilized in both monocular and, progressively, binocular ADAS applications. Front oriented frameworks include in long-extend capacities, for example, path keeping help, cross-traffic alarm, and traffic sign acknowledgment frameworks. Back confronting cameras have delighted in across the board selection basically as a turning around help for the driver. An identical representation perspective on the zone behind the vehicle is shown on a dashboard-mounted screen, now and again expanded with positional designs comparative with controlling wheel development to give leaving direction. Front aligned Binocular or sound system cameras are a later advancement. A couple of cameras can introduce a basically 3D picture that gives the data important to ascertain complex profundity data, for example, the separation to a moving article, making them appropriate for dynamic voyage control and forward impact notice applications.

5 Challenges

Numerous industry players are attempting to improve singular sensors. They are additionally endeavoring to upgrade framework execution through better sensor combination—the intelligible blend of information from various sensors. On the equipment side, intersensory correspondence is a significant test since it requires high transmission capacity and answers for forestalling system over-burdens.

Players are at present improving the parceling and circulation of framework design to address this issue. On the product side, the combination of picture and nonimage information is especially testing. Some OEMs and level one providers are cooperating with the scholarly community to address this test, as can be found in Daimler's coordinated effort with the Karlsruhe Institute of Technology and the University of Ulm.

The restricted usefulness of the present sensors, joined with their significant expense, might be the best limitation to ADAS take-up. Numerous organizations are gaining ground on the two fronts, in any case. As one model, Mobileye and different new businesses are attempting to improve the usefulness of camera-based arrangements, which regularly experience issues recognizing impediments during rainstorms or in different circumstances when perceivability is restricted. In the event that camera-based arrangements make up for lost time to radar and Lidar in usefulness, they could in the long run rule the ADAS advertise in view of their lower cost. "One box" arrangements that consolidate lasers and cameras may likewise become famous in light of the fact that they are more affordable than radar or Lidar alone. This is a significant improvement, since specialists accept that semiautonomous driving will not become a reality until the business has a practical Lidar framework that is completely incorporated with different sensors.

Programming calculations. Running on ECUs and MCUs, calculations utilize the contribution from sensors to integrate the earth encompassing a vehicle progressively (going well beyond the preparing that sensors have just finished). The calculations at that point give yield to the driver or indicate how the framework ought to effectively intercede in vehicle control. This could require the absolute generally complex in-vehicle programming combination at any point made, since any choices that the calculations indicate, for example, the use of crisis brakes, are basic to guaranteeing security.

Because of improvements in sensor combination, the industry is going to progress from inserted programming running on a solitary ADAS-explicit ECU to programming stages running on unified ECUs or MCUs. These product stages have a more elevated level of deliberation to permit adaptable mix of sensor-combination calculations. Industry players are currently concentrating on making such calculations, which consider increasingly precise union of sensor information and progressively productive preparing, in light of the fact that they will help forestall information over-burden or lulls. Another need is making calculations that take into account more secure vehicle route and all the more precisely anticipate all conceivable human conduct—including possibly nonsensical reactions—in different circumstances, for example, when an impact between two autos seems impending.

Mapping. At the point when GPS inclusion comes up short, for example, during burrow travel, itemized and precise mapping frameworks can help forestall mishaps. These frameworks additionally store geological and foundation data, making refreshes varying, and speak with locally available sensors to decide a vehicle's precise area. OEMs and different players in the car business are searching for lower-cost strategies to build and look after maps. The absolute latest arrangements incorporate sending "mapping vehicles" outfitted with 3-D lasers and 360-degree superior quality cameras. Guide engineers are additionally utilizing information from sensors introduced on business armadas, for example, FedEx, just as GPS information from drivers.

Both the USA and the European Union have given decisions commanding that by year 2020 all vehicles must be outfitted with self-ruling crisis stopping mechanisms and forward-crash notice frameworks. These life-sparing choices alone would be sufficient to make ADAS (Advanced Driver Assistance Systems) the principle focal point of consideration for some, particularly makers, OEM merchants and their providers. Furthermore, increasingly more vehicle purchasers are getting delicate to ADAS applications and highlights which would improve economy and solace, such as leaving help or vulnerable sides observing. To wrap things up, completely self-sufficient vehicles will be showcased not all that numerous years from now and the enormous business capability of this open door is clear not exclusively to vehicle makers, yet in addition to cutting edge organizations like Google and other, who are spending gigantic entirety on innovative work to create attractive arrangements in oneself driving autos commercial center.

An ongoing McKinsey report says that, however, present incomes for ADAS sellers are as yet moderate, most industry specialists hope to see a yearly increment of in excess of 10% from 2015 to 2020. Development paces of this levels would be among the most noteworthy that the car division and related ventures had ever known (Fig. 1).

ADAS incorporates numerous advancements, as appeared in the picture; the principle ones respect processors, sensors, mapping, and programming calculations. Normally, it is programming calculations which include picture handling and as a result need work by RSIP Vision. We have done that in a few tasks, a large portion of which are recorded in our ADAS segment. Our calculations utilize input originating from sensors to combine continuously chosen data from the encompassing street condition. These calculations are created for two principle purposes: (1) to give yield to driver, as caution or other data; (2) to determine the manner in which framework ought to respond to control the vehicle (brake, steer, or other well-being related route orders).

These calculations are not an insignificant advancement. In the expressions of the report:

> they could require the absolute generally complex in-vehicle programming incorporation at any point made, since any choices that the calculations determine, for example, the use of crisis brakes, are basic to guaranteeing security.

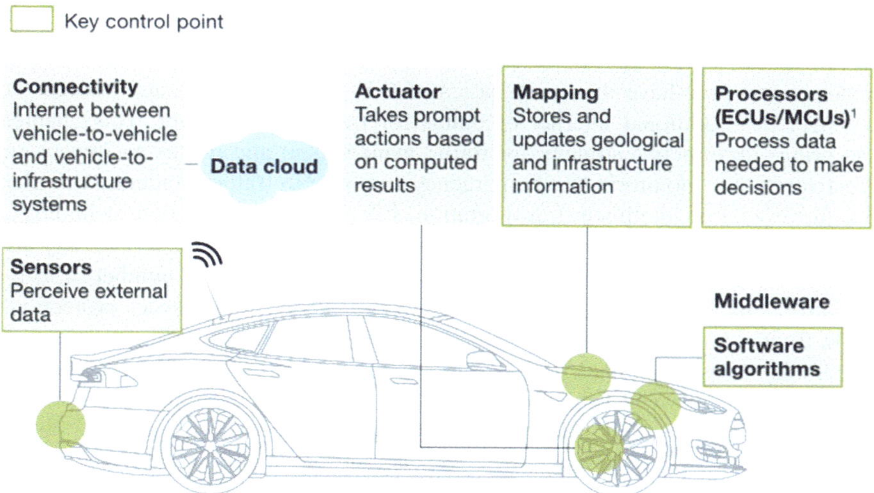

Fig. 1 Technologies integrated in ADAS (http://groupementadas.canalblog.com/archives/2018/04/17/36303306.html)

A portion of these calculations are created to precisely foresee all conceivable human conduct, which is by definition very eccentric and possibly nonsensical. This is critical to permit more secure vehicle route when, for example, an accident between two vehicles or with a person on foot seems up and coming to the ADAS application.

6 Challenges of Radar-Based ADAS System

Despite the fact that radar-based frameworks have an advantage over vicinity-based frameworks there are many deficiencies of these frameworks.

- Detecting little items with Radar is generally hard for shorter frequencies.
- Static object recognition whose relative speed is actually that of the moving item, acquiring essential perusing is very troublesome and requires exclusive calculations to address this issue.
- Existing radar-based ADAS frameworks do not function admirably in shut situations, for example, burrows and for the most part go into backup mode.
- Radar-based frameworks have their constraints with regard to perceiving and arranging objects.
- Interference from other radar framework, which causes precision issue.
- Time taken in dense location and cautioning sign to the driver is on the higher side. This require genuine intercession in lessening handling time with the goal that the waring and other prudent step are taken as fast as would be prudent.

7 Conclusion

In our chapter, we have discussed basics of IoV, its architecture, and concept of IoV over the traditional internet of things for traffic management. This chapter also brings up effective strategy for traffic management and makes an impact on effectiveness for monitoring and emergency response to traffic incidents. In order to achieve goal of intelligent transportation, IoV relies on perception technology, intelligence, computing technology, and various other technologies. In order to truly deepen people's life and rapid development, IoV needs to focus a number of technical problems, for instance, information sharing and collection, privacy protection, etc. In future, development focus of IoV must not be limited to "surfing of internet" but should focus on connection, i.e. on communication. After establishment of communication between vehicles and vehicles, vehicles and people, etc., we can have transfer of information in a truly manner, i.e. without hindrance and hence IoV plays a vital role. In order to have better development and construction of IoV, government departments must play a strong organizational and coordination function. They should also formulate relevant policies and laws support and guide the establishment of individual alliances of IoV.

References

1. Yang, F., Wang, S., Li, J., Liu, Z., & Sun, Q. (2014). An overview of internet of vehicles. *China Communications, 11*(10), 1–15.
2. Peng, H., Li, D., Abboud, K., Zhou, H., Zhao, H., Zhuang, W., & Shen, X. (2017). Performance analysis of IEEE 802.11p DCF for multiplatooning communications with autonomous vehicles. *IEEE Transactions on Vehicular Technology, 66*(3), 2485–2498.
3. Xu, W., Omar, H. A., Zhuang, W., & Shen, X. (2017). Delay analysis of in-vehicle internet access via on-road WiFiaccess points. *IEEE Access, 5*(1), 2736–2746.

iDriveAR: In-Vehicle Driver Awareness and Drowsiness Framework Based on Facial Tracking and Augmented Reality

Ahmad Hoirul Basori and Sharaf J. Malebary

1 Introduction

The accident rate at Saudi Arabia is very high and attract significant attention from the government to apply some policy, including the camera implementation in the highway and traffic light junction. The data from Ministry of Interior, general directorate of traffic, the accident in 2017 (1438H) has reached 47,368 cases. The mortality rate around 0.75% and 4.15% has caused injury [1]. At the beginning of 2000, most of the accident case caused by over speeding or disobeying the traffic signal [2]. However, this trend has changed and become more dangers these days with the booming of mobile phones usage. It mentioned that over 160.000 cases caused by using a mobile phone while driving [3]. The use of a mobile phone might lead the user to be seriously injured or even mortality. Therefore, this paper presenting an innovative solution by augmenting driver awareness and prevent drowsiness by giving alert to the driver. Even though there is a lot of previous research has been done, our unique solution relied on the depth image which is acquired from a depth camera that has more precision toward motion tracking. The structure of the paper consists of five main parts: first segment report presenting the introduction of the research. In contrast, the second segment focused on the related works of current or previous research. The third segment reports the methodology of the study, and it is continued by result and discussion of the proposed approach. Finally, the conclusion is presented as the closing of the paper along with the future direction of the research.

A. H. Basori (✉) · S. J. Malebary
Faculty of Computing and Information Technology Rabigh, King Abdulaziz University, Rabigh, Makkah, Kingdom of Saudi Arabia
e-mail: abasori@kau.edu.sa

© Springer Nature Switzerland AG 2021
N. Gupta et al. (eds.), *Internet of Vehicles and its Applications in Autonomous Driving*, Unmanned System Technologies, https://doi.org/10.1007/978-3-030-46335-9_7

2 Problem Analysis and Related Works

Facial expression becomes a basis of drowsiness detection. The previous researcher took the driver's face and track its feature then classify using pattern recognition technique [4]. It is found that the facial expression has a strong correlation with brain signal condition when human start to feel drowsy [4]. The other research conducted a pilot test a pilot test in the military setup has been made to test the blink occurrence that affects driver focus, and attention as an effect of drowsiness then give alert to them [5]. Naurois et al. use artificial neural network (ANN) as an algorithm to improve the drowsiness detection by implanting the knowledge of trained user to the newbie [6]. They were able to predict when the drowsiness will happen in the future so that the system can trigger an alert. The other researcher gives a quite high prediction with 96% accuracy for micro-sleep up to 5 min in advance [7, 8]. The pedestrian accident also become a concern of researcher. Therefore, a model of pedestrian collision warning systems is proposed based on head-up display using augmented reality technology that will give alert information in front of the driver screen [9]. Instead of drowsiness, the other researcher also considers braking behavior as an essential factor to reduce the accident rate. They focused on transition time toward Brake Pedal Force and Brake to Maximum Brake, which is conducted on driving simulation [10]. Time pressure that occurred because of traffic jam, lateness, or other factors could trigger the driver to apply brake aggressively to avoid accident [11]. The emotional state such as an anger and impatience along with time constraint are considered as factors that might lead the time constraint is considered as factors that might lead them to crash or collision [12–14]. The driver behaviour at traffic light need to be evaluated, because many accident are triggered by running on red light also need to be monitored because of the crash because of passing traffic light too high [15]. The other mobile application also has the potential to be used as trackers for driver awareness and give them alert whenever distraction is occurred. Most of the previous work focused on in-vehicle awareness which is the driver as a focused point. At the same time, the proposed method consists of two main cameras which are tracking the driver awareness, and it is incorporated with second cameras that monitored the object in front of the car.

3 Research Method and Material

This section will reveal the material that we used during experiment. The material that used for development can be detailed out as follow:

- Kinect Depth Camera (2 units)
- HD Logitech Camera
- PC

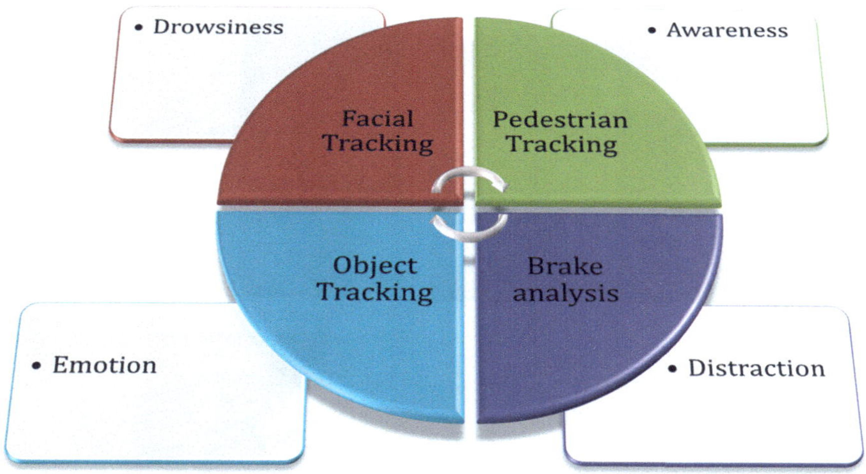

Fig. 1 Proposed framework for iDriveAR

Figure 1 describes the proposed system architecture that consists of several components:

- Accident cause: drowsiness, awareness, distraction, emotion
- Preventive action: facial tracking, pedestrian tracking, object tracking, brake analysis

All factors mentioned in Fig. 1 are intercorrelated each other; emotion may lead to drowsiness, awareness, and distraction, while facial tracking and brake analysis focused on to analyze the user behavior. Pedestrian and object tracking is more toward the object in front of the car. The initial setup for our experiment is shown in Fig. 2.

Figure 2 portrays the position of the initial setup for two primary cameras, Kinect depth camera used for facial tracking to detect the driver awareness and drowsiness. It is installed in front of the driver and on the passenger side to give two different angles of view. The detail specification of the camera is depicted in Table 1.

4 Result and Discussion

The system is initiated by tracking the facial features of the driver to obtain the desired behavior. The general process of face tracking is accomplished by conducting the detection and location of faces within the same period or initiate detection then location. Afterwards, a tracking algorithm is called to handle features of the face that will be changed when the driver moves their faces [16] (Fig. 3).

Fig. 2 Experimental setup

Table 1 Camera specification

No.	Name/types	Specification
1	Kinect camera	• Depth sensor coverage: 1.2–3.5 m • Depth image stream (speed): 320 × 240 16-bit, 30 fps • Color image stream: 640 × 480 32-bit, 30 fps • Audio stream (desert): 16-bit, 16 kHz • Field of view – Horizontal: 27° – Vertical: 43° – Motor tilt range – Vertical: ±27°
2	Logitech HD camera—C922 pro stream	• FullHD:1080p@30fps/720p@60fps for streaming and can be operated continously • The field of view (FoV): 78°s • It consists of glass protection and advanced autofocus • Cable length: 5 ft • Power: USB port

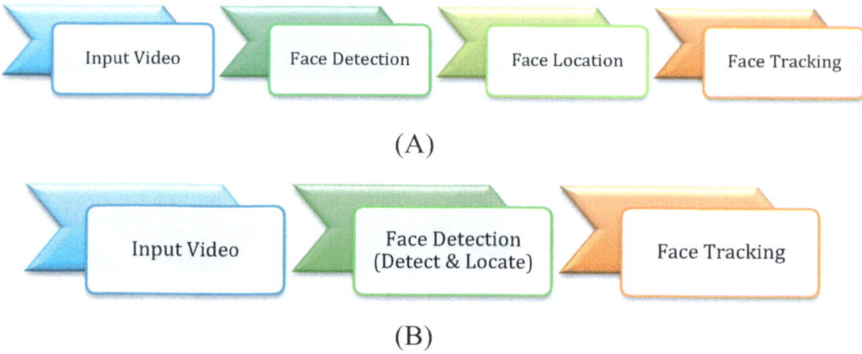

Fig. 3 Face detection phases

Afterwards, the recognition process involves neural network approach with fuzzification. The fuzzy partition function has two class membership i and j which is illustrated in Eq. 1, where $\phi(xk)\ k = 1, \ldots, p$ represent set of vector. d_i represents the distance from the vector inside class I, while c manages the rate with fuzzy membership.

$$\varphi = 0.5\frac{e^{c(d_j - i)/d} - e^{-c}}{2\left(e^{c} - e^{-c}\right)} \tag{1}$$

$$\varphi = 1 - \varphi_i\left(x_k\right) \tag{2}$$

The experiment is conducted in a controlled situation where safety of driving is our main concern.

Figure 4 illustrates the position of driver's face when its tracked by a camera, it is shown that driver still steady and focused toward its journey. In Fig. 5, the facial features of driver face such as eye area, nose, and lips got tracked.

The subsequent experiment is to detect the movement of driving when looking at into another direction. Figure 5b portrayed that driver moved his face toward the left side, while in Fig. 5c driver facing down toward the steering.

The third camera is to generate an effect of augmented reality. The camera is facing straight along with windscreen. The camera can track pedestrian or object in front of the car, such as a traffic symbol. Figure 6 illustrate the movement of pedestrian that is moving and Fig. 7 illustrates how the person is going tracked by the proposed system.

To prove the proposed concept, four machine learning techniques involved in the recognition parts as shown in Fig. 8 with Tree as the highest recognition rate around 87% followed by KNN and neural network, while Naïve Bayes has the lowest recognition rate on this case.

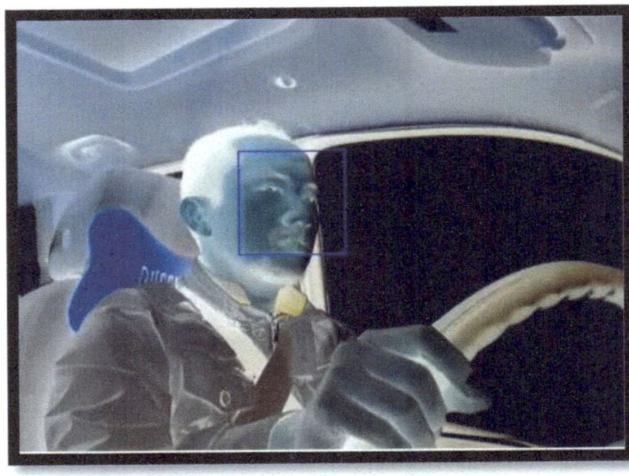

Fig. 4 Face detection

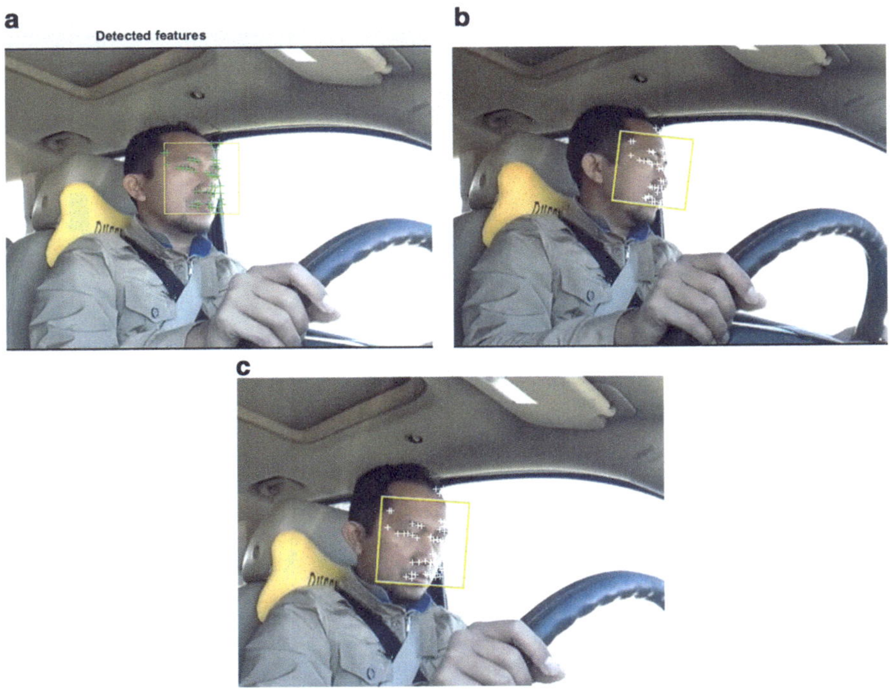

Fig. 5 (**a**) Facial features tracking. (**b**) Facial tracking for looking to the left side. (**c**) Facial tracking for facing down

In addition to the evaluation part, Table 2 contains a value for ROC analysis that is given to measure acceptability of the proposed approach that evaluated through several machine learning algorithms.

Fig. 6 Pedestrian situation

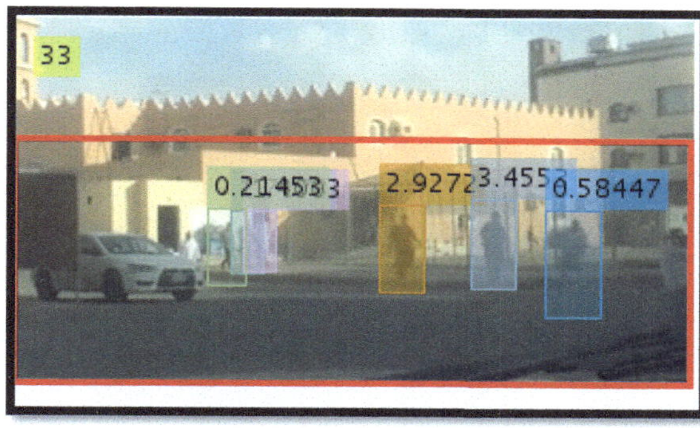

Fig. 7 Pedestrian tracking

Evaluation Results

Model	AUC	CA	F1	Precision	Recall
kNN	0.819	0.771	0.755	0.759	0.771
Tree	0.954	0.873	0.867	0.864	0.873
Neural Network	0.768	0.729	0.696	0.714	0.729
Naive Bayes	0.620	0.322	0.422	0.692	0.322

Fig. 8 Evaluation result

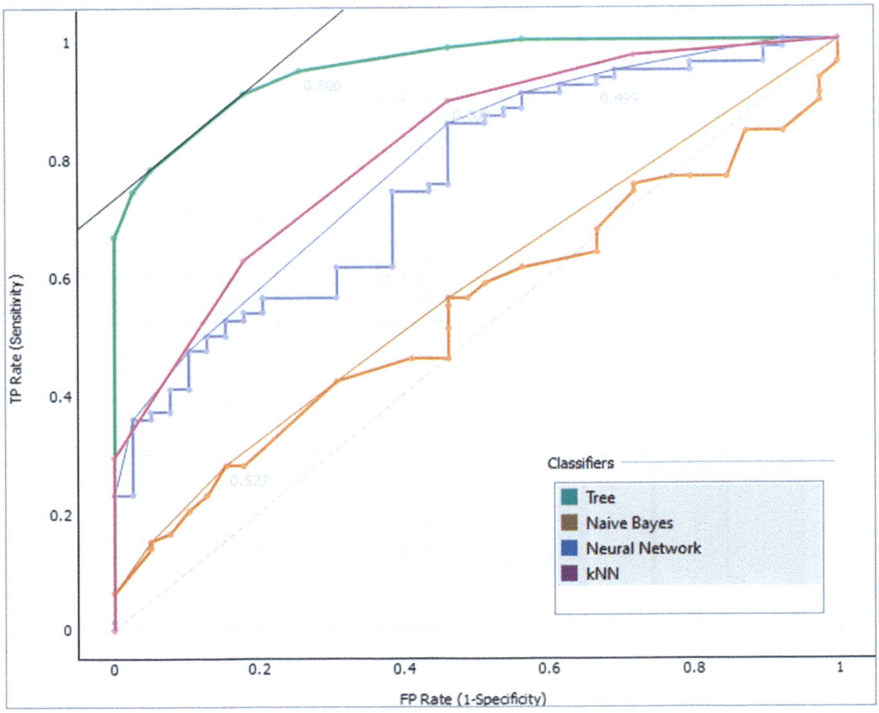

Fig. 9 ROC curve analysis

$$\text{TP Rate} = \frac{\text{TP}}{\text{TP} + \text{FN}} \tag{3}$$

$$\text{FP Rate} = \frac{\text{FP}}{\text{TN} + \text{FP}} \tag{4}$$

where TP Rate: True Positive Rate and FP Rate: False Positive Rate. TP: True Positive, FN: False Negative, FP: False Positive.

Figure 9 shows that sensitivity is increased significantly when the specificity is increased, based on several repeated experiment it is found that the proposed tracking approach is acceptable and able to detect the driver awareness and pedestrian motion during the test. The TP Rate related to sensitivity, while FP Rate focused on specificity. As we can see from the Fig. 9 that Tree algorithm, which is represented by the green curve, shows its superiority among other tree algorithm and located far from the curve threshold.

Table 2 ROC analysis value for ROC curve analysis

	Driver types	Tree	Naïve Bayes	Neural network	kNN	Tree (D)	Tree (A)	Naïve Bayes (D)	Naïve Bayes (A)	Neural network (D)	Neural network (A)	kNN (D)	kNN (A)
1	D	D	D?	D	D	1.000	0.000	0.297	0.164	0.732	0.263	1.000	0.000
2	D	A	A	A	D	0.250	0.750	0.246	0.437	0.460	0.530	0.600	0.400
4	D	D	D?	D	D	0.500	0.500	0.182	0.117	0.667	0.315	0.600	0.400
5	D	A	D?	D	D	0.000	1.000	0.070	0.048	0.499	0.490	0.600	0.400
6	D	A	D?	A	D	0.000	1.000	0.096	0.189	0.327	0.647	0.600	0.400
7	D	D	D?	D	A	0.960	0.000	0.066	0.031	0.682	0.285	0.400	0.600
8	D	D	D?	D	A	0.960	0.000	0.083	0.051	0.578	0.386	0.400	0.600
11	D	D	D	D	D	0.750	0.250	0.550	0.023	0.903	0.083	0.600	0.400
13	D	D	D?	D	D	1.000	0.000	0.288	0.223	0.714	0.280	1.000	0.000
14	D	A	D?	D	A	0.000	1.000	0.288	0.223	0.697	0.294	0.400	0.600
16	D	D	D?	D	D	1.000	0.000	0.036	0.024	0.854	0.144	0.600	0.400
17	D	A	D?	D	A	0.250	0.750	0.068	0.049	0.498	0.492	0.200	0.600
20	D	D	D?	D	D	0.857	0.000	0.319	0.199	0.800	0.156	0.800	0.200
21	D	D	A	D	D	1.000	0.000	0.292	0.460	0.554	0.420	0.600	0.400
23	D	A	D	D	D	0.000	1.000	0.517	0.034	0.802	0.195	0.600	0.400
24	D	A	D	D	D	0.000	1.000	0.471	0.198	0.644	0.354	0.600	0.400
25	D	D	D?	D	D	1.000	0.000	0.329	0.177	0.732	0.262	0.800	0.200
26	D	D	D?	D	D	0.500	0.500	0.103	0.027	0.937	0.063	1.000	0.000
31	D	A	A	A	D	0.000	1.000	0.139	0.661	0.376	0.582	0.600	0.400

5 Conclusion and Future Works

Driver focus and awareness for sure is a key in safety driving. Therefore a lot of effort was being used to enhance the safety driving. Overspeeding and braking habits might cause crash or collision on the road. This research focused on providing an alternative solution by putting the camera inside the car and analyzed the driving behavior. The camera which facing the driver can track the driver motion during the journey. While the third camera can identify the pedestrian and provide a proximity alert toward the driver, when the distance is very close, a warning message will be triggered to alert the driver. The evaluation analysis shown that Tree algorithm capable to achieve 87% recognition rate followed by KNN algorithm. We realized that there are still a lot of works to do to improve the tracking features and result. However, as overall observation, camera trackers on the driver and road will assist the driver toward another level of awareness.

Acknowledgement This work was supported by the Deanship of Scientific Research (DSR), King Abdulaziz University, Jeddah Saudi Arabia. The authors, therefore, gratefully acknowledge the DSR technical and financial support.

References

1. Ministry of Interior,General Directorate of Traffic. (2017). Traffic accident statistic. Retrieved January 2020, from https://data.gov.sa/Data/en/dataset/traffic-accident-statistics-as-of-1438-h
2. Ansari, S., Akhdar, F., Mandoorah, M., & Moutaery, K. (2000). Causes and effects of road traffic accidents in Saudi Arabia. *Public Health, 114*(1), 37–39.
3. Nabbout, M. (2019). Over 160,000 car accidents in Saudi Arabia are caused by phone usage.Retrieved January 2020, from https://stepfeed.com/over-160-000-car-accidents-in-saudi-arabia-are-caused-by-phone-usage-9800
4. Hachisuka, S., Ishida, K., Enya, T., & Kamijo, M. (2011). Facial expression measurement for detecting driver drowsiness. In D. Harris (Ed.), *Engineering psychology and cognitive ergonomics* (pp. 135–144). Heidelberg: Springer.
5. Aidman, E., Chadunow, C., Johnson, K., & Reece, J. (2015). Real-time driver drowsiness feedback improves driver alertness and self-reported driving performance. *Accident Analysis and Prevention, 81*, 8–13.
6. de Naurois, C. J., Bourdin, C., Bougard, C., & Vercher, J.-L. (2018). Adapting artificial neural networks to a specific driver enhances detection and prediction of drowsiness. *Accident Analysis and Prevention, 121*, 118–128.
7. Watson, A.,&Zhou, G.(2016). Microsleep prediction using an EKG capable heart rate monitor. *2016 IEEE First International Conference on Connected Health: Applications, Systems and Engineering Technologies* (*CHASE*). 328–329. https://doi.org/10.1109/CHASE.2016.30
8. Galarza, E. E., Egas, F. D., Silva, F. M., Velasco, P. M., & Galarza, E. D. (2018). Real time driver drowsiness detection based on driver's face image behavior using a system of human computer interaction implemented in a smartphone. *Advances in Intelligent Systems and Computing, 721*, 563–572. https://doi.org/10.1007/978-3-319-73450-7_53.
9. Phan, M. T., Thouvenin, I., &Frémont, V. (2016). Enhancing the driver awareness of pedestrian using augmented reality cues.*The 19th IEEE International Conference on Intelligent Transportation Systems* (*ITSC*), Windsor Oceanico Hotel, Rio de Janeiro, Brazil, November 1–4, 2016.

10. Pawar, N. M., Khanuja, R. K., Choudhary, P., & Velaga, N. R. (2020). Modelling braking behaviour and accident probability of drivers under increasing time pressure conditions. *Accident Analysis and Prevention, 136*, 1–17.
11. Rendon-Velez, E., van Leeuwen, P. M., Happee, R., Horváth, I., van der Vegte, W. F., & de Winter, J. C. F. (2016). The effects of time pressure on driver performance and physiological activity: A driving simulator study. *Transportation Research Part F: Traffic Psychology and Behaviour, 41*, 150–169. https://doi.org/10.1016/j.trf.2016.06.013.
12. Paschalidis, E., Choudhury, C. F., & Hess, S. (2019). Combining driving simulator and physiological sensor data in a latent variable model to incorporate the effect of stress in car-following behaviour. *Analytic Methods in Accident Research, 22*, 100089. https://doi.org/10.1016/j.amar.2019.02.001.
13. Techer, F., Ojeda, L., Barat, D., Marteau, J., Rampillon, F., Feron, S., & Dogan, E. (2019). Anger and highly automated driving in urban areas :The role of time pressure. *Transportation Research Part F: Traffic Psychology and Behaviour, 64*, 353–360. https://doi.org/10.1016/j.trf.2019.05.016.
14. Naveteur, J., Cœugnet, S., Charron, C., & Dorn, L. (2013). Impatience and time pressure:Subjective reactions of drivers in situations forcing them to stop their car in the road. *Transportation Research Part F: Traffic Psychology and Behaviour, 18*, 58–71. https://doi.org/10.1016/j.trf.2012.
15. Palat, B., & Delhomme, P. (2016). A simulator study of factors influencing drivers' behavior at traffic lights. *Transportation Research Part F: Traffic Psychology and Behaviour, 37*, 107–118. https://doi.org/10.1016/j.trf.2015.11.009.
16. Kasheevnik, A., Lashkov, I., & Gurtov, A. (2019). Methodology and mobile application for driver behavior analysis and accident prevention. *IEEE Transactions on Intelligent Transportation Systems, 1*, 1–10. https://doi.org/10.1109/TITS.2019.2918328.

The Concept of Fusion for Clear Vision of Hazy Roads in ADAS

M. Dhana Lakshmi Bhavani and R. Murugan

1 Advanced Driver Assistance Systems

Advanced driver assistance systems, or ADAS, is the fast-growing advancement in vehicle technology which is designed to improve the safety to passengers sitting in the car as well as the pedestrians crossing the road. Such systems are very helpful in reducing the probability of motor vehicle accidents. The driver assistance system is embedded with lot of controlling units which can warn passengers of sudden dangers, and control is passed immediately to avoid the severity or prevent from accidents. In simple words, an ADAS is designed to compensate the human mistakes, non-activeness, and erroneous control while driving a vehicle [1].

In simple view, ADAS or *Advanced Driver Assistance Systems* are designed to assist the manual driver in the process of driving the vehicles, maintaining the safe drive, comfortable driving experience [2]. There are various modes of operations available in ADAS; some are built-in traits and others are available as add-on features. Afterwards, it is going to be *Automatic Driver Assistance System*. Such systems should be undergone multiple verifications and expanding the other innovative technologies to improve the each and every assistance system. The need for such an expansion of innovatory driving ideas are to manage a comfortable and convenient journey for passengers. A computerized arrangement of ADAS system in the vehicles are advised to reduce the causalities of road accidents and loss of life due to human errors. As the technology advances, the manufacturing companies are taking more responsibility for maintaining the self-calibrating sensors to reduce the complexity of overall assistance unit. Additionally, there are aftermarket solutions

---

M. D. L. Bhavani (✉) · R. Murugan
Department of Electronics & Communication Engineering, National Institute of Technology Silchar, Silchar, Assam, India
e-mail: murugan.rmn@ece.nits.ac.in

© Springer Nature Switzerland AG 2021

N. Gupta et al. (eds.), *Internet of Vehicles and its Applications in Autonomous Driving*, Unmanned System Technologies, https://doi.org/10.1007/978-3-030-46335-9_8

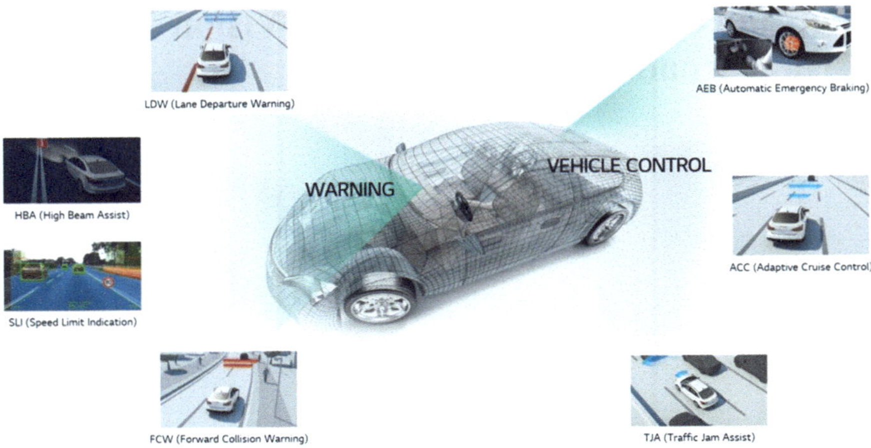

Fig. 1 Assistance and control systems for ADAS

available. Some of the warning and vehicle control units are shown in Fig. 1. These systems include:

- Lane Departure Warning (LDW),
- Forward Collision Warning (FCW),
- High Beam Safety System (HSS),
- Traffic Signals Recognition (TSR),
- Adaptive Cruise Control (ACC),
- Traction control,
- Automatic Emergency Break (AEB),
- Speed Limit Indication (SLI),
- Pedestrian Crash Avoidance Mitigation (PCAM) and more.

Now-a-days, the ADAS systems are well-developed and increasingly playing a vital role in the development of safe vehicle driving assistance system with fewer accidents and resulting in reduced damages. The most important and familiar of all the control assistances is the anti-lock braking system, or ABS. If any obstacle or sudden entry of pedestrian were identified on road, immediate alert passage signal is given to the ABS to stop the car or reduce the actual speed of the car. Various incarnations of ABS have been developed and trialled by automotive manufacturers for better crash avoidance. More safety features should be added to the central electronic control unit to avoid collisions by undertaking smooth control over the vehicle. The completely automated system to the driver assistance must provide most of the key features to reduce road fatalities by avoiding the human errors.

2 Pedestrian Crash Avoidance Mitigation (PCAM)

Pedestrian Crash Avoidance Mitigation (*PCAM*) framework is otherwise referred as Pedestrian Protection or identification module which utilizes a Personal Computer (PC) and an artificial intelligence system to acknowledge the crossing of pedestrians or bicycles in the path of drive and to draw an action for safety. The system of PCAM are frequently embodied for Pre-Collision avoidance module which is already accessible in most of the exclusive automobile manufacturers [3]. According to the current survey, in the USA more than 6500 pedestrians and 900 bicyclists were killed per annum. However, in world more than 3 lac pedestrians were affected each annum. An efficacious framework was developed extensively which could be capable of rescuing more than 40% of lives. The safety of pedestrians is considered to be secondary important role to Passenger safety.

Generally, PCAM frameworks are a part of the innovation in self-driving vehicles, for example, the Tesla Autopilot and utilize an incorporated front aligned camera and radar or Lidar framework intended to help relieve or keep away from a frontal accident. However, the advancements in PCAM systems do not require self-driving innovations, just cameras and radar. Some of the time, these can be upgraded with the additional features like detection of pedestrians and cyclists in poor lighting.

Under specific considerations, if the PCAM frameworks establish that the probability of occurring a front collision with a pedestrian or cyclist is more, it assists the driver to make equivocal move and brake by utilizing a sound and visual caution. If at all the driver perceive the danger and brakes, the system may utilize a type of brake help to give extra braking power. If the driver doesn't apply brake and the PCAM confirm that the danger of crash with a pedestrian or bike is very high, the framework may consequently apply the brakes, and reduce the rate of collision. Most of the time, this is a setting the driver must make to prior initiation, however it is very default part [4].

Different control units are embedded within the assistance system to monitor the surroundings of the vehicles as well as the neighbouring vehicles to it. All these units are programmed to avoid the crashes, one such important control unit to avoid the loss of life and property is *Pedestrian Crash Avoidance Mitigation* (*PCAM*). A scenario pedestrian crossing the road of practical situation in general is as shown in Fig. 2.

The latest advancements in computational systems avail us to recognize the pedestrian with the concept of pattern recognition innovation in the field of artificial intelligence. Generally, an AI system uses machine learning or deep learning with CNN which make a decision based on millions of scenarios. In a simplified explanation, images from vehicles front camera and radar were matched to the already stored prototypes in the personal computers. If any similarity is found, the immediate acting systems in the PCAM are awaked for further step. Different technologies can be discovered for the PCAM system with furthermore information from the other connected neighbouring cars.

Fig. 2 A scenario of pedestrian crossing the road

This type of PCAM systems is extending their use in Advanced Driver Assistance Systems which can achieve the pedestrian safety. The automobile industry Volvo is the foremost developer of automatic braking system which is focused on neighbouring cars and also the pedestrians. With such an innovation, there is more than 60% improvement in automatic pre-collision detection with the automatic braking system.

3 Automatic PCAM on Hazy Roads

The Pedestrian Crash Avoidance Mitigation (PCAM) control unit helps to identify the persons crossing the road in front of the cars and if there is any such situation arises, immediate signal is forwarded to the central controlling assistance system to slow down the speed of the car. The corresponding crash avoidance system is tested well for its implementation but there are some inevitable situations due to environmental conditions. The climate changes based on season manipulate the humidity content in atmosphere. Winter is one of the states of difficulty while driving on roads. Heavy fog or Haziness in the atmosphere causes the loss of clear vision of roads.

Multiple hazardous situations may arise for vehicle drivers while driving in a heavy fog condition. A few discoveries have uncovered that in instances of different auto crashes particularly those including chain responses, haze quite often has an influence in such an incident [5]. According to our survey reports, many fatalities and severe injuries are occurring every year due to haze-related collisions. The studies revealed that, with fog/haze/smoke-related crashes, more automobiles are involved and the injury severity if much greater than that of the clear-visibility crashes [6, 7]. Smoke is certainly not an immediate result of climate or precipitation, yet typically an outcome of human dealings. The effect of smoke on collisions, notwithstanding the impacts of haze, has been examined in past work. Smoke-related perceivability situations are rare and have a significantly unexpected quality in comparison to times of decreased perceivability brought about by precipitation and dampness, (for example, haze or rain) in light of the fact that there is no water on the windshield nor need to utilize wipers. The effect of harsh climate conditions on car accidents has been a significant street well-being concern. Proof has set up the relationship among decreased perceivability under foggy or cloudy climate conditions, crash rate and damage seriousness dependent on authentic accident information. For example, the danger of a lethal accident expanded with the presence of haze and smoke and the probability of serious damage crash under foggy climate conditions changed with geographic area.

An equivalent situation creates a typical confusion while driving, because the road is visible only up to few metres. The corresponding control unit of the ADAS may be muddled. Such uncertain situations should be overcome with the advancements in the latest technologies. An example of a pedestrian crossing the heavy hazy roads in the morning is given in Fig. 3. From the figure, we can also observe that the traffic indication signal on the right top is also not so enough to view. A camera system or a sensor system is already embedded on the front side of the vehicle to continuously monitor the scene. But during winter morning days, the sunlight is not so good up to few hours, so that the normal sensing camera is not so advisable since it cannot remove the haziness in the scenes. So, a suggestible device to minimize the collisions in winter season are *infrared radiation (IR) thermal imaging camera systems*. The infrared light is having large wavelengths than the visible light camera systems. The light absorption and emission by infrared radiation create a critical effect on earth's climate. The natural IR image cannot detect the haziness or fog in scenes. So, this type of feature is used as an advantage in our typical problem.

Numerous circumstances call for cameras that can enter through haze and cloudiness. These cameras can picture day or night, dispensing with the exorbitant requirement for various cameras covering a similar scene. This is particularly significant in virtual control towers, airborne surveillance, railroads, fringes, and port and harbour security situations where haze and fog are a customary event. Near infrared and multispectral cameras help with the location and recognizable proof of suspicious targets, tourist spots and different objects of intrigue, potential coasting items, or moving toward ships on the water.

Fig. 3 Pedestrian crossing on Hazy road

Even in clear skies, innate environmental retention places restrain on how far a specific infrared camera can see. A thermal imaging camera delivers a picture dependent on the distinctions in thermal radiation that an article discharge. Basically, for the more distance objects this infrared signal needs to withdraw from the objective to the camera, a greater amount of that signal can be lost along the drive. Accordingly, the attenuation component should be considered. This is the proportion of the incident radiation to the radiation transmitted through a protecting material. Damp air goes about as a "shield" for infrared radiation. Summer month environments for the most part have a higher constriction contrasted with winter a very long time because of expanded stickiness levels. In this way, accepting clear skies and great climate conditions, you will have the option to see more remote with a warm imaging camera in winter than in summer.

But here the humidity in air is only an example of how we lost the infrared radiation. Other than the humidity there may be various climatic conditions which can detriment more to the range of a thermal imaging camera. Fog and rainfall can strictly constraint the scope of a thermal imaging system due to scattering of light off droplets of water. The increase in complexity of thickness of droplets, the more the infrared signal is weakened. An imperative query that clients may rise is what is the range of rain or fog will can constraint the performance of a thermal imaging infrared system, and how does this compare to the range in the visible region of the spectrum.

Fog is a visible cumulative of small sized water drops suspended in the environment at or near the surface of earth. When air is almost saturated with water vapour, this means that the relative humidity is close to 100%, and that fog can form

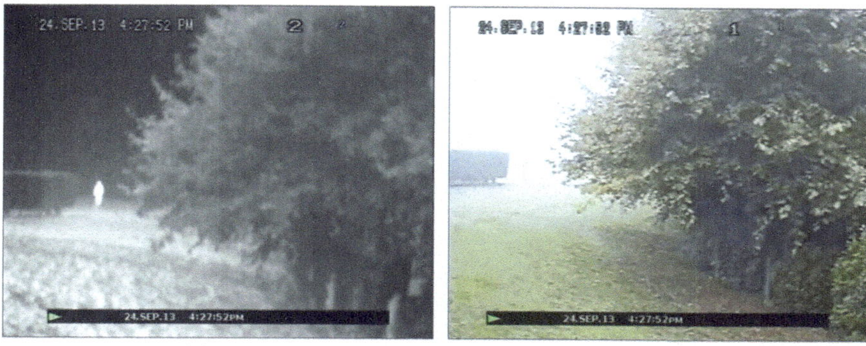

(a) (b)

Fig. 4 (**a**) Infrared image. (**b**) Visible image

in the presence of a sufficient number of condensation nuclei, which can be smoke or dust particles. There are various categories in fog. Advection fog is accumulated through the combination of two different masses through dissimilar temperatures and moisture. One more type is radiative fog. This is established during the radiative freezing of the air at temperatures near the dew point. Few of the fog banks are thicker than the other because the water drops may grow more through accretion. In foggy condition, water drops can absorb more water and grows the size. Depending on the size of water drops structure, we can decide whether the scattering is less in the IR waveband compared to the visible range or not. The cause for dreadful conditions of visibility in a foggy environment is the absorption and scattering of natural or artificial illumination of foggy particles. The extent of absorption and scattering changes on the microphysical structure of the minute foggy parts, which are also called aerosols [21, 22].

From the above figure, we can identify the presence of a person on road from Fig. 4a but in visible image of same scenario, in Fig. 4b, is not showing the existence of person due to climate moisture or fog.

4 The Concept of Fusion

Fusion is a concept of combining features of multiple files to get an effective information. The obtained information from the fused file is more accurate, valuable, and useful than from the individual image. Here we can concentrate on fusing the multimedia files like images. For example, if we have different images of same scenario with various lightening conditions. Let us have an image with different resolutions as shown in Fig. 5. Figure 5a gives a low-resolution image, whereas Fig. 5b gives a high-resolution image. By investigating these two image files, the pixel

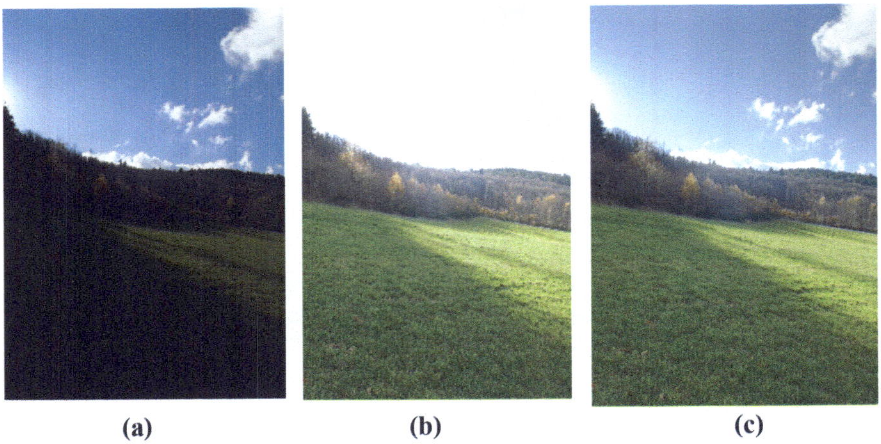

Fig. 5 (**a**) Low-resolution image, (**b**) high-resolution image, and (**c**) fused image

information is not completely clear in both the cases. After performing the operation of fusion, Fig. 5c gives the good visual appealing than the first two images.

By comparing the information gathered by a single sensor image and multi-sensor images can provide more information with feature complementarity for complex and complete scene representations with improved visual perceptual quality. By gathering corresponding information from different sensor systems, we can produce a single composite image, which is more instructive and appropriate for further proceedings and decision-making. In particular, infrared and visible image fusion is an indispensable branch that plays a vital role in military and civilian applications and has been used to improve the performance in terms of human visual perception, object detection, and target recognition [8].

Visible measuring devices can acquire reflected light to provide background details, such as vegetation, texture, area, and soil with high spatial resolution because they operate in the same waveband that can be recognized by human vision. However, under non-satisfying lightening conditions, original information may be lost. By contrast, infrared sensors map thermal radiation emitted from objects to grey images; hence, these sensors demonstrate distinctive benefits in overcoming adverse light conditions and highlighting thermal targets, such as pedestrians and vehicles [9]. Nevertheless, due to drawbacks of hardware and environments, infrared images are often accompanied by blurred details, serious noise, and considerably low resolution. These restrictions inspire the fusion of infrared and visible images to provide a good quality image, which can concurrently emphasize the thermal radiating conditions for target identification and reserve texture information for characterization of appearance.

Affording to numerous interpretations and applications, the process of fusion can be performed at the following three levels: pixel, feature, and symbol levels. Pixel-level fusion is on the lowest level and fuses different physical parameters.

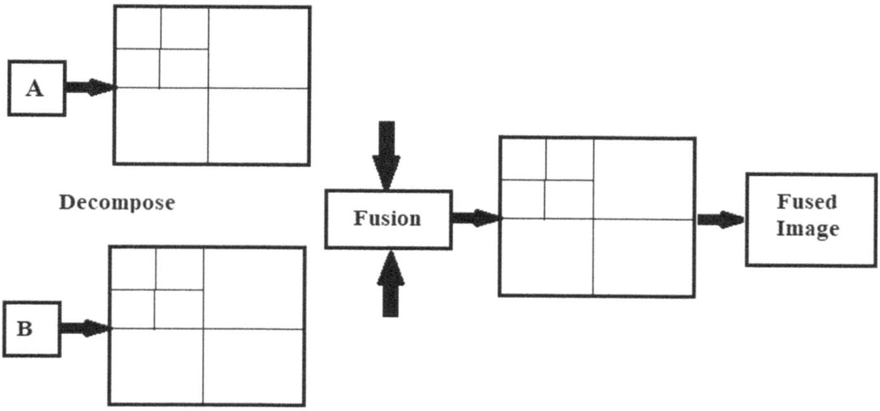

Fig. 6 The concept of fusing two different images of same scenario

Feature-level fusion involves feature descriptor, probability variables, and object labels. Symbol-level fusion uses high abstraction of information, such as representative symbols or decisions, for fusion.

The functional scenario of fusing two images is as given in Fig. 6. Both the images are decomposed using the concept of discrete wavelet transform with wavelet basis function. After decomposing the images, extracting the best picture features from two selected images and then combining the features to get an accurate quality of image. Finally applying inverse discrete wavelet transform for normal visualization of image.

5 Standard Image Fusion Methods

Categorizing the image fusion approaches affording to the information entering the fusion and according to the principle of performing fusion. We categorize the following methods.

- *Multiview fusion* of images from the same modality and taken at the same time but from different viewpoints.
- *Multimodal fusion* of images coming from different sensors (visible and infrared, CT, and NMR, or panchromatic and multispectral satellite images).
- *Multitemporal fusion* of images taken at different times in order to detect changes between them or to synthesize realistic images of objects which were not photographed in a desired time.
- *Multi-focus fusion* of images of a 3D scene taken repeatedly with various focal lengths.

- *Fusion for image restoration*. Fusing two or more images of the same scene and modality, each of them blurred and noisy, may lead to a deblurred and denoised image. Multichannel deconvolution is a typical representative of this category.

This approach can be extended to super resolution fusion, where input blurred images of low spatial resolution are fused to provide us a high-resolution image. In each category, the fusion consists of two basic stages: image registration, which brings the input images to spatial alignment, and combining the image functions (intensities, colours, etc.) in the area of frame overlap. Image registration works usually in four steps.

- *Feature detection*. Salient and distinctive objects (corners, line intersections, edges, contours, closed boundary regions, etc.) are manually or, preferably, automatically detected. For further processing, these features can be represented by their point representatives (distinctive points, line endings, centre of gravity), called in the literature control points.
- *Feature matching*. In this step, the correspondence between the features detected in the sensed image and those detected in the reference image is established. Various feature descriptors and similarity measures along with spatial relationships among the features are used for that purpose.
- *Transform model estimation*. The type and parameters of the so-called mapping functions, aligning the sensed image with the reference image, are estimated. The parameters of the mapping functions are computed by means of the established feature correspondence.
- *Image resampling and transformation*. The sensed image is transformed by means of the mapping functions. Image values in non-integer coordinates are estimated by an appropriate interpolation technique.

We present a survey of traditional and up-to-date registration and fusion methods and demonstrate their performance by practical experiments from various application areas. Special attention is paid to fusion for image restoration, because this group is extremely important for producers and users of low-resolution imaging devices such as mobile phones, camcorders, web cameras, and security and surveillance cameras. Image fusion methods can be broadly classified into two groups—spatial domain fusion and transform domain fusion.

1. *Pixel-level fusion*: Pixel-based fusion is performed on a pixel-by-pixel basis. It generates a fused image in which information associated with each pixel is determined from a set of pixels in source images to improve the performance of image processing tasks such as segmentation.
2. *Feature-level fusion*: Feature-based fusion at feature level requires an extraction of objects recognized in various data sources. It requires extraction of salient features which are depending on their environment such as pixel intensities, edges or textures. These similar features from input images are fused.
3. *Decision-level fusion*: Decision-based fusion consists of merging information at a higher level of abstraction, combines the results from multiple algorithms to yield a final fused decision. Input images are processed individually for information

extraction. The obtained information is then combined by applying decision rules to reinforce common interpretation.

After an in-depth survey, the thesis presented in [9] identifies that selection of a suitable fusion level depends on available information type. For example, when the sensors are alike, fusion at pixel level can be opted since it considers all the data into account. But they produce spatial distortions in the fused image. Feature level fusion can be opted when the features found while processing the information from different sensors can be appropriately associated. When sensors are very different, fusion at decision level will be more suitable and is also computationally more efficient. In the present study, different block-based feature-level image fusion techniques are identified and given to the neural network for better classification purpose.

Various fusion principles such as averaging, Bravery method, principal component analysis (PCA), and IHS based approaches come under spatial domain methodologies. Other imperative spatial domain fusion method is the high pass filtering-based technique [11–15]. Here the high frequency details are injected into up-sampled version of MS images. The limitation of spatial domain approaches is that they produce spatial distortion in the fused image. Spectral distortion becomes a negative factor while we go for further processing, such as classification problem. Spatial distortion can be very well handled by frequency domain methodologies on image fusion. The multiresolution analysis has become a very useful tool for analysing remote sensing images. The discrete wavelet transform has become a very useful tool for fusion. Some other fusion methods are also there, such as Laplacian pyramid based, curvelet transform based, etc. These methods perform well in spatial and spectral quality of the fused images compared to other spatial domain-based approaches of fusion.

The images used in image fusion should already be registered. Misregistration is a major source of error in image fusion. Some well-known image fusion methods are:

- High pass filtering technique
- IHS transform-based image fusion
- PCA based image fusion
- Wavelet transform image fusion
- Pair-wise spatial frequency matching

Despite the notable progress of image fusion, fusing infrared and visible images of different resolutions remain a challenging task. Infrared images constantly suffer from considerably lower resolution compared with visible images. Under this condition, if existing methods are applied to fusing source images, then eliminating resolution differences, through down-sampling visible images or up-sampling infrared images as an example, is the first problem to solve. While performing the operation of fusion on multiple images of same scenario there is much redundant data since all the images are same. So, such redundant data should be nullified while performing the operation of fusion.

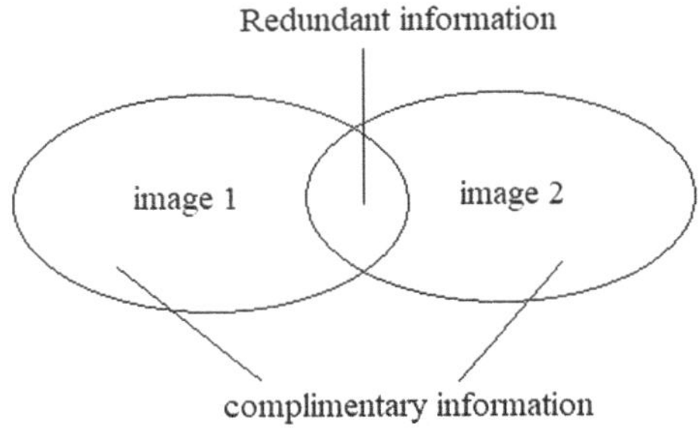

Fig. 7 Basic scheme for image fusion without redundant information

Figure 7 shows the basic image fusion scheme. In this figure image 1 and image 2 are combined to extract the information between those two images into a single image. This figure consists of the two sections namely redundant information and complimentary information. The reliability of image is improved by redundant information and the capability of image is improved by complimentary information.

6 A Survey on Fusing Visible and Infrared Images

The concept of image fusion is an important method which concentrates on combining the multiple images from various sensors to obtain the robust or more informative image which can avail us to process the image for making decisions. The key role for an admirable fusion methodology is an effective image feature extraction and suitable fusion methods, which allow useful information to be extracted from input images and combined in the fused image without presenting any artefact in the procedure. In the age of information explosion, sensor techniques are being developed rapidly. The appearance of complicated applications requires comprehensive information about a certain scenario for enhanced understanding of various conditions. Sensors of the similar type acquire information from only one aspect and are thus unable to produce all essential information. Hence, fusion methodologies play an increasingly significant role in advanced applications and computer vision [6, 8].

Images of different types, such as visible, infrared, computed tomography (CT), and magnetic resonance imaging (MRI), are good source images for fusion. Among the combinations of these types, infrared and visible image fusion is superior in many aspects. First, their signals come from different modalities, thereby providing scene information from different aspects; i.e., visible images capture reflected light, whereas infrared images capture thermal radiation [16]. Therefore, this combination is more informative than that of single-modality signals. Second, infrared and visible images present characteristics that are inherent in nearly all objects. Furthermore, these images can be acquired by relatively simpler structure, as opposed to such imaging techniques as CT and MRI, which have rigorous requirements. Finally, infrared and visible images share complementary properties, thus producing robust and informative fused images. Visible images typically have high spatial resolution and considerable detail and chiaroscuro; thus, they are suitable for human visual perception. Therefore, these images can be easily influenced by severe conditions, such as poor illumination, fog, and other effects of bad weather. Meanwhile, infrared images, which depict the thermal radiation of objects, are resistant to these disturbances but typically have low resolution and poor texture.

Fusion techniques for visible and infrared images can thus be used in more fields than other fusion schemes due to the ubiquitous and complementary characteristics of the utilized images. Object recognition, detection, image enhancement, surveillance, and remote sensing are typical applications of infrared and visible image fusion. Fusion algorithms can be developed with different schemes and generally divided into seven categories according to their adopted theories, i.e., multi-scale transform, sparse representation, neural network, subspace, and saliency-based methods, hybrid models, and other methods. Multi-scale transform-based methods are the most active field in image fusion and assume images to be represented by various layers in different grains. These methods decompose source images into several levels, fuse corresponding layers with particular rules, and reconstruct the target images accordingly. Popular transforms used for decomposition and reconstruction include wavelet, pyramid, curvelet, and their revised versions. Sparse representation-based methods work on the basis of the possible representation of images with linear combinations of sparse bases in over-complete dictionaries, which are key to their good performance. Neural network-based methods imitate the perception behaviour of the human brain to deal with neural information; this procedure has the advantages of good adaptability, fault tolerance, and anti-noise capacity. Subspace-based methods use similar ideas but implement them in spaces with complete bases instead of over-complete ones. Principal component analysis (PCA), non-negative matrix factorization (NMF), and independent component analysis (ICA) are the main methods in this category. Saliency-based methods are based on the fact that human visual attention is often captured by objects or pixels that are more significant than their neighbours. In addition, these approaches can retain the integrity of salient object regions and improve the visual quality of the fused images. The abovementioned infrared and visible image fusion methods have their advantages and disadvantages. Meanwhile, hybrid models combine the advantages of these schemes to improve image fusion performance. The other techniques are based on total variation, fuzzy theory, and entropy. These methods can lead to a new vision for infrared and visible image fusion, which can inspire the design of novel models [17].

The research on fusion algorithms in theory and application has been rapidly developing in recent years. However, certain problems require attention. An appropriate information extraction method is key to ensuring good fusion performance. Several current approaches, especially multi-scale transform-based methods, depend on predefined transforms and corresponding levels for decomposition and reconstruction. However, no criteria are used to evaluate these transforms and levels. Consequently, transforms are selected blindly, thereby easily degrading performance. Moreover, current fusion rules are mostly excessively simple that certain artefacts, such as halo, are introduced in the results. Methods in other areas also suffer from several drawbacks such as constructing suitable over-complete dictionaries that have good representation of target data using sparse representation-based methods is still difficult. In addition to that, suitable sparse coding techniques are also important for the performance of fusion. Designing appropriate neural networks and adjusting the corresponding parameters remain challenging. Furthermore, applying state-of-the-art deep learning techniques to infrared and visible image fusion is yet to be addressed. Finding a subspace with powerful expressive ability in subspace-based methods is difficult. Meanwhile, designing niche targeting and anti-noise saliency detection methods in accordance with the human visual system is also challenging, and using the saliency of source images is still an open problem. In addition, combining the advantages of various methods to create enhanced hybrid image fusion models is a prominent issue in infrared and visible image fusion. Designing new and good infrared and visible image fusion methods with novel ideas is also difficult. Infrared and visible image fusion is useful, but its current methods still struggle with several problems. Thus, this review concentrates to specifically and comprehensively review the existing fusion methodologies of infrared and visible images and hopes to shed light on the improvement of such approaches. This work has certain advantages in comparison with previous reviews. The current study finds that infrared and visible image fusion is considerably superior to other integration of input images. However, current aims are mainly about general fusion approaches and pay only a small amount of attention to methods for infrared and visible image fusion, thus lacking pertinence [18]. This survey is comprehensive and specific, reviewing, and concluding nearly all main approaches for infrared and visible image fusion. Furthermore, this work includes theories and applications in infrared and visible image fusion and discusses corresponding tendency in future study. Finally, extensive experiments are conducted in this survey to provide an objective comparison of the typical methods and support relative engineering with credible and solid evidence.

Using this concept of fusion, the effect of haziness in the image can be reduced to get a clear vision of roads in order to avoid the pedestrian crash in winter morning season. So, using two camera systems to continuously capture the visible and IR images of same scene. After capturing, the images are forwarded to the fusion system to get the best features from both the images. Finally, the road with clear vision is obtained as a result [19, 20]. Figure 8 shows the same scene with different frequency bands. Figure 8a is a visible camera image and Fig. 8b is an infrared camera image.

7 Performance Metrics

As the number of image fusion techniques are increasing, there is a growing need for metrics. In recent years, a number of computational image fusion quality assessment metrics have therefore been proposed. The goals of the quantitative measures are

Fig. 8 Images of same scenario. (a) Visible camera. (b) IR camera

normally used for the result of visual inspection due to the limitations of human eyes. The following optimal measures are implemented to judge the performance of fusion methods as follows:

7.1 Peak Signal-to-Noise Ratio (PSNR)

PSNR is an engineering term for the ratio between the maximum possible power of a signal and the power of corrupting noise that affects the fidelity of its representation. Because many signals have a very wide dynamic range, PSNR is usually expressed in terms of the logarithmic decibel scale. The PSNR is most commonly used to measure the quality of reconstructed images.

7.2 Mean Square Error (MSE)

Mean square error is a measure of image quality index. The large value of mean square means that image is a poor quality.

7.3 Entropy

Entropy is a parameter to evaluate the information quantity contained in an image. The maximum value of entropy informs that the fused image is better than the reference images.

References

1. Adella, E., Varhelyia, A., & Della Fontana, M. (2011). The effects of a driver assistance system for safe speed and safe distance e a real-life field study. *Transportation Research Part C: Emerging Technologies, 19*(1), 145–155.
2. van Arem, B., van Driel, C. J. G., & Visser, R. (2006). The impact of cooperative adaptive cruise control on traffic-flow characteristics. *IEEE Transactions on Intelligent Transportation Systems, 7*(4), 429–436.
3. Coelingh, E., & Eidehall, A. (2010). Collision warning with full auto brake and pedestrian detection e a practical example of automatic emergency braking. In *13th International IEEE Conference on Intelligent Transportation Systems*, Madeira Island, Portugal, pp. 155–160.
4. Wang, X., Wang, M., & Li, W. (2014). Scene-specific pedestrian detection for static video surveillance. *IEEE Transactions on Pattern Analysis and Machine Intelligence, 36*(2), 361–374.
5. Tarel, J.-P., Hautière, N., Cord, A., Gruyer, D., & Halmaoui, H. (2010). Improved visibility of road scene images under heterogeneous fog. In *IEEE Intelligent Vehicles Symposium*, San Diego, CA, pp. 478–485.
6. Calderara, S., Piccinini, P., & Cucchiara, R. (2011). Vision based smoke detection system using image energy and color information. *Machine Vision and Applications, 22*(4), 705–719.
7. Narasimhan, S. G., & Nayar, S. K. (2003). Contrast restoration of weather degraded images. *IEEE Transactions on Pattern Analysis and Machine Learning, 25*(6), 713–724.
8. Jian, L., Yang, X., Zhou, Z., Zhou, K., & Liu, K. (2018). Multi-Scale image fusion through rolling guidance filter. *Future Generation Computer Systems, 83*, 310–325.
9. Mustafa, H. T., Yang, J., & Zareapoor, M. (2019). Multi-Scale convolutional neural network for multi-focus image fusion. *Image and Vision Computing, 85*, 26–35.
10. Fattal, R. (2008). Single image dehazing. *ACM Transactions on Graphics, 27*(3), 72.
11. Zhang, B., & Zhao, J. (2017). Hardware implementation for real-time haze removal. *IEEE Transactions on Very Large Scale Integration (VLSI) Systems, 25*(3), 1188–1192.
12. Kumar, R., Kaushik, B. K., & Balasubramanian, R. (2017). FPGA implementation of image dehazing algorithm for real time applications. *Proceedings of SPIE, 10396*, 1039633.
13. Shen, H., Meng, X., & Zhang, L. (2016). An integrated framework for the spatio-temporal spectral fusion of remote sensing images. *IEEE Transactions on Geoscience and Remote Sensing, 54*(12), 7135–7148.
14. Wang, L., Du, J., Zhu, S., Fan, D., & Lee, J. (2016). New region-based image fusion scheme using the discrete wavelet frame transform. In *IEEE International Conference on Intelligent Control and Automation (WCICA)*, pp. 3066–3070.
15. Nirmala, D. E., Paul, A. B. S., & Vaidehi, V. (2013). Improving independent component analysis using support vector machines for multimodal image fusion. *Journal of Computer Science, 9*, 1117–1132.
16. Berman, D., Treibitz, T., & Avidan, S. (2016). Non-local image dehazing. In *IEEE Conference on Computer Vision and Pattern Recognition*, pp. 1674–1682.
17. Narasimhan, S. G., & Nayar, S. K. (2000). Chromatic framework for vision in bad weather. In *Proceedings of the IEEE Conference on Computer Vision and Pattern Recognition. (CVPR)*, Hilton Head Island, SC, pp. 598–605.
18. Choi, L. K., You, J., & Bovik, A. C. (2015). Referenceless prediction of perceptual fog density and perceptual image defogging. *IEEE Transactions on Image Processing, 24*(11), 3888–3901.
19. Zhu, Q., Mai, J., & Shao, L. (2015). A fast single image haze removal algorithm using color attenuation prior. *IEEE Transactions on Image Processing, 24*(11), 3522–3533.
20. Gibson, K. B., Vo, D. T., & Nguyen, T. Q. (2012). An investigation of dehazing effects on image and video coding. *IEEE Transactions on Image Processing, 21*(2), 662–673.
21. Kumar, R., Kaushik, B. K., & Balasubramanian, R. (2019). Multispectral transmission map fusion method and architecture for image dehazing. In *IEEE Transactions on Very Large Scale Integration (VLSI) Systems*, pp. 1063–8210.
22. Tarel, J. P., & Hautiere, N. (2009). Fast visibility restoration from a single colour or gray level image. In *IEEE International Conference on Computer Vision*, pp. 2201–2208.

Part IV
Evaluation Methods and Security Issues

Blockchain-Enabled Security and Privacy for Internet-of-Vehicles

Ferheen Ayaz, Zhengguo Sheng, Daxin Tian, and Victor C. M. Leung

1 Introduction

The proliferation of Internet-of-Things (IoT) have revolutionised the concept of conventional Vehicular Ad-Hoc Networks (VANETs) into Internet-of-Vehicles (IoV), ensuring an overall connectivity of a vehicle with not only other vehicles on road but also with every other object including passengers' smart phones, external management platform, navigation systems and other road users, thereby forming an Intelligent transportation system (ITS). The potential benefits of IoV include increased passengers' safety, smooth traffic flow and availability of infotainment. Also, with an increase in vehicle density on road over the past few decades, high risk of accidents and traffic jams appear very common in daily transport and thus seriously threat road safety [1]. Safe driving conditions can be achieved by implementing IoV in which a vehicle or other connected device can initiate timely and efficient message dissemination in emergency situations, such as accidents [2]. The oncoming vehicles may re-route themselves to avoid traffic congestion by receiving an alert prior to entering into an affected area. However, IoV encounters many challenges in implementation of an efficient message dissemination solution because the vehicles behave differently than other wireless nodes. Due to the

F. Ayaz (✉) · Z. Sheng
Department of Engineering and Design, University of Sussex, Brighton, UK
e-mail: f.ayaz@sussex.ac.uk

D. Tian
School of Transportation Science and Engineering, Beihang University, Beijing, China

V. C. M. Leung
Department of Electrical and Computer Engineering, University of British Columbia (UBC), Vancouver, BC, Canada

© Springer Nature Switzerland AG 2021

N. Gupta et al. (eds.), *Internet of Vehicles and its Applications in Autonomous Driving*, Unmanned System Technologies, https://doi.org/10.1007/978-3-030-46335-9_9

dynamic characteristics and complexity of vehicular networks, IoV possesses some specific characteristics and demands to look after some particular challenges discussed below.

1.1 Characteristics and Challenges of IoV

The communication environment in IoV is variable. For example, a motorway is open and allows one direction movement of vehicles with high mobility. A lane change or a brake event must be communicated timely to the concerned vehicles in order to avoid collision. Therefore, a lightweight networking protocol is required. Multiple vehicles sending the same message may result in packet collision and broadcast storm. The connected vehicles must be able to select appropriate relay nodes among themselves, which can effectively forward a message to a large number of vehicles [3]. A central node like a Road Side Unit (RSU) can be used for message dissemination. However, it requires a large investment for installation [4] and its volume and placement positions are crucial for system performance [5]. Moreover, RSU assisted routing protocol may result in higher latency because of increased route discovery time and hops per route [6]. Since low latency is an important criteria when propagating messages to fast moving vehicles, a decentralised network in which vehicles can select relay nodes among themselves independently without any central authority (CA) is desired as a more scalable and cost-effective approach than the networks dependent upon RSU [7].

On the other hand, a residential area contains random movements and obstacles like tall buildings with the presence of relatively larger number of connected devices. An incident occurring in a street cannot be directly witnessed by passengers travelling on an adjacent road. In such situations, it is also essential for vehicles in IoV to authenticate a message before dissemination to maintain trust and security in the network. A malicious vehicle may generate a false message about an incident which did not actually occur on road or deliberately mark an authentic message as fake [8]. Moreover, due to increasing privacy concerns, the identity of a vehicle must not be revealed if it sends a message. However, anonymous announcements can affect the credibility of a message [9]. Public key management by CA is one of the solutions to privacy but large scale networks in dense areas would result in running of several cryptographic operation requests at the same time which is not a feasible solution [10].

Although message dissemination, trust management, privacy and security in connected vehicles are widely discussed in literature [11–13], there is a need to develop an integrated approach in which efficient and trusted communications of IoV can be managed in a decentralised fashion. Furthermore, to maintain the sustainability of message dissemination in IoV, economic modelling is needed which results in an efficient incentive distribution strategy to encourage positive cooperation and punish negative behaviour of vehicles. There are two common types of incentive strategies: price based and reputation based. In price based strategy, messages are treated as commodities which are exchanged for virtual credits [14, 15]. On the contrary, reputation based strategies use measurement of trustworthiness to enforce cooperation. A threshold of reputation is set to distinguish a node behaviour as malicious or non-malicious. A punishment scheme is usually applied for malicious behaviour [16, 17]. Game theoretical analysis [18] suggests that an integrated strategy including the advantages of both price based and reputation based schemes is more effective in promoting cooperation and detecting malicious behaviour in Mobile Ad-hoc Networks (MANETs) [19].

1.2 Blockchain as a Potential Solution

In this chapter, we briefly explain the concept of a blockchain-enabled IoV to overcome the discussed challenges. A blockchain is basically a peer-to-peer electronic cash system in which transaction data is maintained in a ledger as immutable timestamped blocks and each block is linked to the previous block with the help of a cryptographic hash [20]. To complete a transaction in a blockchain, a client needs to first submit a proposal which is then verified by a network of peers using an algorithm known as consensus. If a proposal is verified, it is stored in the blockchain as a unique record. A blockchain is encrypted, decentralised, secure and immutable. Therefore, it is able to solve issues related to security and privacy, and has the potential to contribute towards IoT applications [21, 22]. Our chapter shows how blockchain can be used to implement distributed, secure and privacy-preserving communications in IoV. The goal is to introduce a secure, lightweight and decentralised peer-to-peer vehicular network in which untrusted vehicles can interact with each other in a verifiable manner. The key challenge in IoV is ensuring the authenticity of message in a decentralised manner. It is interesting to point out that the feature of consensus in blockchain can be applied to resolve this challenge. Blockchain provides robustness against dissemination of false messages as long as the honest vehicles collectively possess more controlling power than malicious vehicles. The basic idea of blockchain implementation for security and privacy in IoV is shown in Fig. 1. A blockchain-enabled IoV would be able to record messages

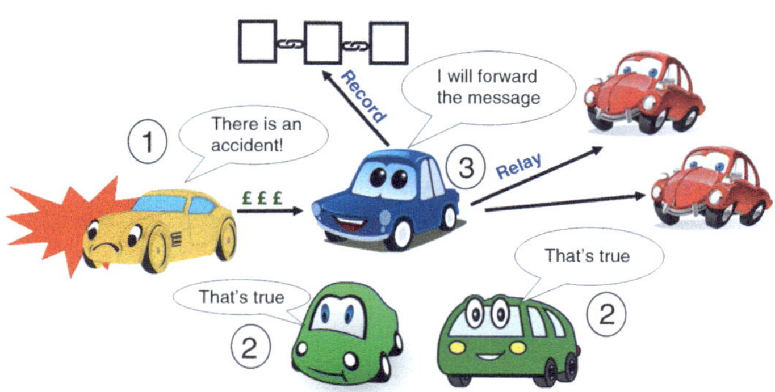

Fig. 1 Basic idea of blockchain for security in IoV

into an ongoing chain along the broadcast route, forming a ledger that cannot be changed. All vehicles can collectively act as peers to authenticate messages and select a relay node to forward message and manage the history of publicly available previous record of messages. To discourage accidents, ensure smooth traffic and promote positive cooperation, the vehicles causing accident can compensate by paying reward to the vehicles participating in message dissemination.

2 Fundamentals of Blockchain

This section briefly describes the fundamentals of blockchain, including its structure, types, basic consensus algorithms and their suitability with respect to IoV.

2.1 Structure of Blockchain

As shown in Fig. 2, a blockchain is used to store transactions in a sequence of blocks chained together by including a cryptographic link to the previous block known as hash [23]. In a conventional blockchain, whenever a new transaction is made, it is announced across the network. Nodes are able to verify a transaction and record the verified transaction into a block containing an encrypted hash of previous block, thereby making it cryptographically secure. Nodes which append a block to the blockchain after verification and broadcast it in the network are known as miners. Miners have to go through a mutual agreement called consensus algorithm to create a block which makes it fraud-proof [24]. All nodes in a network update their copy of blockchain regularly in order to ensure consistency in the entire distributed ledger [25].

A typical blockchain is permissionless and open to public where anyone in a network can make entry into the ledger. Bitcoin and Ethereum are the examples of permissionless blockchain [23]. On the other hand, Hyperledger family offers various versions of permissioned blockchain, where specific roles and access

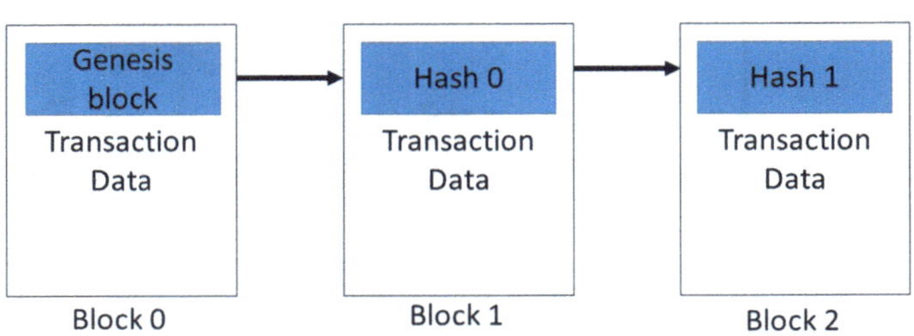

Fig. 2 The blockchain structure

limitation can be associated to certain nodes. Considering different roles in ITS, such as automotive manufacturers, government and transportation authorities, a permissioned blockchain is considered to be appropriate for vehicular applications [26].

2.2 Consensus Algorithm

The core of a blockchain mechanism is consensus algorithm. It is an agreement to validate a transaction and unanimously take a decision. The Bitcoin uses an incentive based Proof-of-Work (PoW) consensus for block-mining. In PoW, the nodes have to solve an extremely difficult puzzle to become a miner, which requires high computational power and results in large propagation delay. However, a difficult cryptographic puzzle which cannot be solved and tampered easily makes the blockchain highly secure. An alternative of PoW is Proof-of-Stake (PoS) [27]. New versions of Ethereum are PoS-based [28]. PoS replaces the mining operation of PoW with a miner selection on the basis of highest amount of stakes or virtual currency owned, which makes it unfair for members possessing smaller stakes [29]. A permissioned blockchain usually adopts voting based Practical Byzantine Fault Tolerant (PBFT) algorithm for achieving consensus among a group of nodes in a permissioned environment [30]. PBFT consists of three stages as shown in Fig. 3. In first stage, a client sends a transaction proposal to validating peers. In second stage, a transaction proposal is validated by receiving votes from at least a certain minimum number of peers. In third stage, the block is committed to the blockchain. The minimum number of votes required to validate a transaction proposal is usually specified in a smart contract of a blockchain. A novel PBFT based voting algorithm known as Yet Another Consensus (YAC) is implemented in Hyperledger Iroha blockchain platform. It tolerates f faulty nodes out of $2f+1$ participants taking part in consensus [31]. A comparison of basic blockchain technologies is summarised in Table 1, which indicates that consensus algorithm of Hyperledger; PBFT is the most promising algorithm for vehicular applications, as it overcomes the limitations of PoW and PoS. However, PBFT does not scale well with large number of nodes, as the number of messages broadcasted to validate a transaction proposal increases with raising number of nodes. Nevertheless, with moving vehicles, the number of nodes within a limited transmission range of a vehicle initiating a transaction

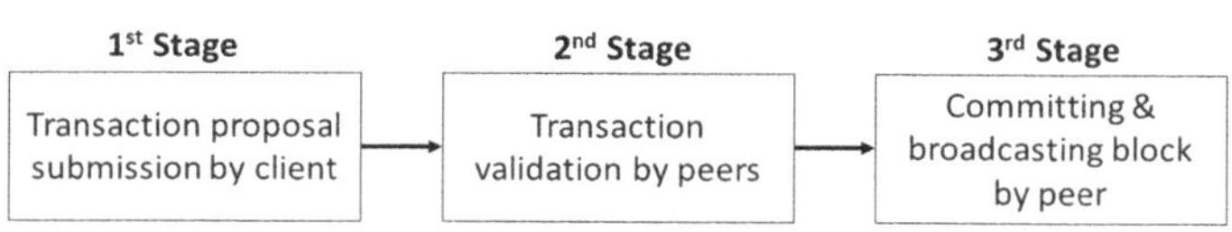

Fig. 3 PBFT consensus algorithm

Table 1 Comparison of blockchain technologies

	Bitcoin	Ethereum	Hyperledger
Consensus	PoW	PoS	PBFT
Mode	Permissionless	Permissionless	Permissioned
Currency	Bitcoin	Ether	Not required
Smart contract	No	Yes	Yes
Pros	Highly secure	Time and energy saving	Time and energy saving
Cons	Time and energy consuming	Biased towards peers with high amount of stakes	Message count increases with number of nodes

proposal is always limited and cannot be increased to a large extent. Therefore, despite the drawback of PBFT, it is well-suited to be used with IoV.

3 Related Work

In this section, the related work which utilises blockchain technology as a potential solution to existing challenges in IoT and IoV applications is discussed with referenced literature. The discussion is focussed on how blockchain can the existing challenges of privacy and trust in IoT and IoV.

3.1 Privacy

The transactions in a blockchain are recorded with the use of cryptographic public and private keypairs. The public key can also be changed for different transactions. The anonymity in network is maintained by keeping the identity of participants private. This features makes blockchain useful in many applications which include confidentiality, for example, storing patients' record and history [32]. A permissioned blockchain solution to keep the privacy of identity and location of electric vehicles during payment at charging stations was proposed in [33]. Due to lack of scalability in permissioned blockchain's consensus, public blockchain for urban areas with large number of vehicles was also suggested in [33]. It can be deduced that greater privacy can be achieved through permissioned blockchain by trading off scalability. Apart from payment transactions, the privacy-preserving feature of blockchain can also be used to ensure anonymous message exchange. In [34], a blockchain-based incentive anonymous announcement network was proposed, which used virtual credits, known as, CreditCoin as a reward for forwarding the message. The transactions of CreditCoins were managed by blockchain, whereas vehicle-to-vehicle (V2V) communications were implemented separately.

The privacy of vehicles was maintained because their identities were encrypted during communications. The message dissemination system was not completely decentralised because it relied on RSUs or official public vehicles as servers which undertake the work of consensus. In [35], the concept of blockchain managed by a Certificate Authority was proposed, which assigned or revoked encrypted keys. It provided a privacy-preserving solution because vehicles communicated with each other using public key instead of revealing their original identities. For reputation management, the vehicles needed to report to a centralised law enforcement agency whenever they observed a malicious behaviour.

3.2 Trust and Reputation

In a blockchain network, nodes have to go through a consensus algorithm to verify a transaction in a block. A consensus algorithm ensures trust in a network because fake transactions cannot be verified by all nodes [36]. By use of blockchain in IoV, a malicious vehicle cannot easily initiate a false message, because it has to be propagated after validation. One way to judge the validity of a message is on the basis of reputation of sender. This is why reputation management is one of the subjects where blockchain research is rapidly progressing. Many researches have employed reputation to reach consensus because PoW is not feasible to be used with moving vehicles. A blockchain-based reputation management with the consensus algorithm of PoS, where reputation was the stake of vehicles was proposed in [37]. The vehicles reported their reputation opinions of other vehicles to a nearby RSU. To carry out consensus algorithms, the reputation values were downloaded from RSU and a mining vehicle was elected on the basis of highest reputation. The vehicle was considered malicious if its reputation value was below a certain threshold. A similar approach was presented by Yang et al. [38] where RSUs stored reputation of vehicles and maintained blockchain. To carry out consensus algorithm, mining RSUs underwent a joint PoW and PoS algorithm where reputation was regarded as a stake. It also considered Bayesian Inference for authentication of incidents occurring on road. Both mechanisms relied upon RSUs for blockchain management. Consensus processing by RSUs to store trust ratings of vehicles in a blockchain was also proposed in [39]. Blockchain managed by RSUs was recommended in [40] because of their higher storage capacity and processing power than vehicles. However, the issues of short connection time between a RSU and a vehicle as well as security attacks on RSUs which control blockchain are still not addressed in existing literature. In [41], permissioned blockchain of reputation, not dependent upon RSUs, was proposed and messages were authenticated on the basis of sender's reputation. It followed conventional PoW algorithm for miner election.

3.3 Decentralisation

The blockchain methodology is independent of third party to finalise transactions. This feature makes it suitable for distributed networks, for example, supply chain [42] and IoV [43]. In [44], authors discussed the strategies to design decentralised applications using blockchain and pointed out that the challenges in developing blockchain-based processes including selection of blockchain network type (permissioned or permissionless) and consensus model. Decentralisation in IoV by the use of Ethereum blockchain was proposed by Leiding et al. [45]. It was suggested to automatically punish negative behaviour of vehicles, for example, over-speeding and ignoring traffic lights, by imposing fine or recalculate tax or insurance by the use of smart contract without requiring a central authority. In [46], an incentive mechanism which was used to motivate vehicles to forward messages was proposed. Only the transactions of incentives were managed by blockchain in a decentralised manner, whereas the relay node selection mechanism was employed separately.

To summarise, it can be stated that the existing literature contains proposed incentive mechanisms for vehicles to participate in message dissemination using blockchain or blockchain-enabled reputation management systems, usually depending upon a central authority or RSUs. However, a consolidated solution can potentially be formulated by involving blockchain not only for transaction of virtual currency but also for transaction of messages to implement completely decentralised, private and trusted IoV communications.

4 Blockchain-Enabled IoV

In this section, the blockchain-enabled solution for message dissemination among connected vehicles, along with reputation management and incentive distribution mechanism [47] is explained. The key notations used in this chapter are defined in Table 2. The proposed solution consists of the following components:

1. **Central Authority** (CA) Before joining the blockchain network, a vehicle needs to be registered with the CA. It assigns a wallet address and a pair of public and private keys to the vehicle for communications and records its original identity. The key pair is used for anonymous transmissions. The role of CA is to grant vehicles an access to a permissioned blockchain system. CA monitors the network and is also in-charge of setting the rules in the system, for example, the minimum amount of balance in a credit wallet and a vehicle's initial reputation value.
2. **Originator** (ORG) An ORG is the vehicle which is involved in an incident and originates a $transaction\ proposal$.
3. **Transaction Proposal** A $transaction\ proposal$ contains an unendorsed message sent by ORG including details about the incident, for example, its location

Table 2 Key notations

Notation	Definition	Notation	Definition
CA	Central authority	ORG	Originator
i	Hop index	$\overline{END}(i)$	Endorsers at ith hop
$RLY(i)$	Relay node at ith hop	TC	Transmission charge
CC	Call compensation paid by ORG	R_j	Reputation of vehicle j
R_T	Reputation threshold	N_{END}	Number of endorsers at hop i
N_{HOP}	Number of hops	Q_j	Quality factor of vehicle j
t_{END}^{max}	Maximum waiting time for *endorsements*	t_{max}	Maximum time limit for message dissemination
df_j	Distance factor of vehicle j	$d_{j,k}$	Distance between vehicle j and k
d_{HOP}^{min}	Minimum distance a message should reach per hop	CQ_j	Channel quality parameter of vehicle j
CP_j	Collision probability of vehicle j	t_W	Time window
N_j^s	Number of successful transmissions by vehicle j	N_j^o	Number of overall transmissions by vehicle j
t_j^{occ}	Channel occupancy time when vehicle j is trying to transmit	$RSSM_j$	Received signal strength matrix of vehicle j
RSS_j	Received signal strength of vehicle j	RSS_T	Received signal strength threshold
G_j^r	Receiving antenna gain of vehicle j	G_j^t	Transmitting antenna gain of vehicle j
TP_j	Transmitting power of vehicle j	λ	Wavelength used in VANET
w_1, w_2	Weight of $\overline{END}$, $RLY(i)$ share in CC	α_1, α_2	Weight of df_j and CQ_j to compute Q_j
TR_j	Transmission range of vehicle j	$P_j(i)$	Profit of vehicle j at hop i
β	Reputation reward	γ	Reputation penalty
ACT_j	Action of player j	U_j	Utility of player j

and time. The voting based blockchain system confirms the authenticity of a *transaction proposal* through *endorsements*.

4. **Endorsement** An *endorsement* is a vote confirming that a *transaction proposal* is authentic. A minimum number of *endorsements*, N_{END}^{min}, is required to consider a *transaction proposal* as an endorsed message.
5. **Endorsers at hop** i**,** $(\overline{END}(i))$ At each hop i, $\overline{END}(i)$ is a set of vehicle j which vote for a suitable relay node. When $i = 0$, $\overline{END}(i)$ are vehicles adjacent to ORG and they also endorse a *transaction proposal*, if they have witnessed the incident itself or made confirmation through camera equipped device or any other service as described in [26].

6. **Relay node at hop** i, ($RLY(i)$) It is the vehicle which further disseminates an endorsed message. It is mutually selected by the voting of $\overline{END}(i)$. It is also one of the $\overline{END}(i)$ and may vote for itself. It also acts as a miner by committing the block to record transfer of virtual credits and reputation updates in the blockchain and finalising the consensus algorithm at each hop i.
7. **Credit Wallet** Each vehicle possesses a credit wallet in which virtual credit is stored.
8. **Transmission Charge** (TC) It is a fee deducted from a vehicle's credit wallet when it originates or endorses a $transaction\ proposal$ or further disseminates an endorsed message. This fee is deposited to CA. CA regulates TC according to the incident recovery cost estimated at the location of incident and time of the day. The purpose of introducing TC is to demotivate vehicles to make a fake $transaction\ proposal$ or $endorsement$, as it is generated at the expense of their virtual credit.
9. **Call compensation** (CC) As a compensation of causing an incident, CC is the amount deducted from the credit wallet of ORG. It is inversely proportional to its reputation. ORG with higher reputation can pay less CC. This is the reason why vehicles are motivated to increase their reputation.
10. **Reputation** (R_j) It is a reputation value of vehicle j. A vehicle is only eligible to endorse a $transaction\ proposal$ if its reputation value exceeds a certain reputation threshold, R_T, that is, $j \epsilon \overline{END}(0)$ if $R_j > R_T$.

4.1 Voting Based Consensus Algorithm

The consensus algorithm of proposed blockchain-enabled IoV follows the voting mechanism of YAC [28] where ORG acts as a client and $\overline{END}(i)$ perform the role of a peer and is illustrated in Fig. 4. When an incident occurs, ORG originates a $transaction\ proposal$. Upon receiving a $transaction\ proposal$, a vehicle j that can confirm the incident becomes an endorser, i.e. $j \epsilon \overline{END}(0)$, broadcasts its cryptographically encrypted signature as a part of $endorsement$ phase and votes for a suitable $RLY(0)$. The selection criteria of $RLY(i)$ are described later in this section. If N_{END}^{min} endorsements are obtained within the time limit, t_{END}^{max}, the $transaction\ proposal$ is considered to be an endorsed message. In order to simplify our assumption, we only consider a static case in which N_{END}^{min} is fixed. An adaptive N_{END}^{min} corresponding to real traffic conditions is out of the scope of this paper, but can be partially solved by using the traffic density estimation method [48]. When a $transaction\ proposal$ is classified as an endorsed message, $RLY(0)$ will further disseminate it in its transmission range and generate a block. The block generation details are described later in this section. Voting based selection of $RLY(i)$ for $i > 0$ is continued until reaching a maximum number of hops, that is, $i > N_{HOP}^{max}$ or the endorsed message has been disseminated until a time limit, d_{max}. It is noted that $\overline{END}(i)$ need to send $endorsements$ for a $transaction$

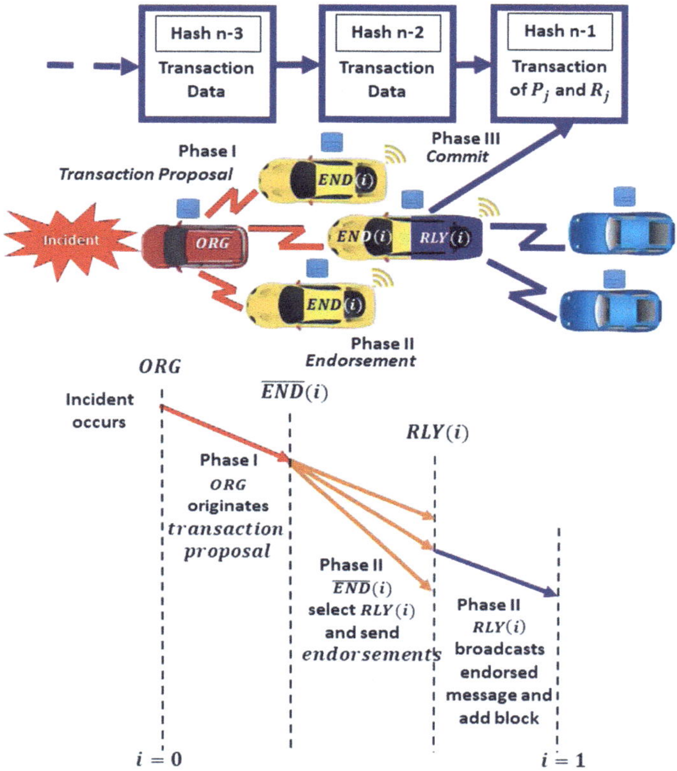

Fig. 4 Proposed voting based consensus algorithm

proposal only when $i = 0$. At $i > 0$, $\overline{END}(i)$ only take part in consensus of $RLY(i)$ selection, because they already receive an endorsed message instead of a *transaction proposal*.

4.2 Relay Selection Mechanism

In our proposed solution, $\overline{END}(i)$ vote for the most appropriate $RLY(i)$ which can further disseminate an endorsed message to a wider area. In this work, we assume that each vehicle j is sharing its location coordinates, channel quality parameter CQ_j, collision probability CP_j, receiving antenna gain G^r_j, transmitting antenna gain G^t_j, maximum transmitting power TP_j and transmission range TR_j during their regular beacon message exchange. The parameters of $RLY(i)$ are stored in the blockchain and are regularly audited by CA to detect and investigate potential fraud if a vehicle cheats by sending fake parameters to achieve highest Q_j. $j \epsilon \overline{END}(i)$

computes the quality factor Q_j and determines its own choice of $RLY(i)$ with the highest Q_j, that is,

$$RLY(i) = index(max(Q_{\overline{END(i)}})), \tag{1}$$

and,[1]

$$Q_j = \alpha_1 df_j + \alpha_2 CQ_j(1 - CP_j) + RSSM_j, \tag{2}$$

where $j \epsilon \overline{END}(i)$, df_j is the distance factor of vehicle j, $RSSM_j$ is the received signal strength matrix, α_1 and α_2 are corresponding weights. df_j is defined as

$$df_j = \begin{cases} \frac{d_{j,ORG}}{d_{HOP}^{min}}, & \text{if } d_{j,ORG} \geq d_{HOP}^{min}, i = 0, \\ \frac{d_{j,RLY(i-1)}}{d_{HOP}^{min}}, & \text{if } d_{j,RLY(i-1)} \geq d_{HOP}^{min}, i > 0, \\ 0, & \text{otherwise,} \end{cases} \tag{3}$$

where d_{HOP}^{min} is the minimum distance a message should be disseminated per hop to restrict $RLY(i)$ selection outside a distance limit [49]. CQ_j and CP_j depend upon internal statistical parameters of medium access control (MAC) as described in [50]. CQ_j is defined as

$$CQ_j = \begin{cases} \frac{N_j^s}{N_j^o}, & \text{if } N_j^o > 0, \\ 0, & \text{otherwise,} \end{cases} \tag{4}$$

where N_j^s is the number of successful transmissions and N_j^o is the number of overall transmissions in a time window. CP_j is estimated as likelihood of collision occurrence if a message is forwarded by vehicle j. It is defined as

$$CP_j = \frac{t_j^{occ}}{t_W}, \tag{5}$$

where t_j^{occ} is the accumulated time at which the channel was occupied or busy when vehicle j was trying to transmit and t_W is a fixed time window. The Received Signal Strength RSS_j, as defined in [51], can be calculated from a distance between locations of vehicle j and j', where $j' \epsilon \overline{END}(i)$, as

$$RSS_j = \frac{G_j^r \times G_j^t \times TP_j}{(4\pi d_{j,j'}/\lambda)^2}, \tag{6}$$

[1] Any other algorithm instead of the proposed method for computation of Q_j can also be used in blockchain-enabled IoV.

where λ is the wavelength used in VANET. The threshold of the received signal strength is defined as

$$RSS_T = \frac{G_j^r \times G_j^t \times TP_j}{(4\pi \times 0.9054TR_j/\lambda)^2}. \tag{7}$$

The range of the antennas is assumed as the circular area of radius TR_j. From [52], it shows that the average distance between two random mobile nodes in a circular region of radius TR_j is $0.9054TR_j$. The purpose of using RSS_T as a threshold parameter is to estimate the reliability of connection with vehicle j. If $RSS_j \geq RSS_T$, the connection can successfully be maintained for a certain time period. Otherwise, the connectivity may be lost before completing a voting consensus [53]. $RSSM_j$ is calculated as

$$RSSM_j = \begin{cases} 1 - \frac{RSS_T}{RSS_j}, & \text{if } RSS_j \geq RSS_T, \\ 0, & RSS_j < RSS_T. \end{cases} \tag{8}$$

4.3 Incentive Distribution Mechanism

The incentive distribution of proposed solution is managed by the blockchain. The proposed voting based blockchain is used to store parameters related to $Q_(RLY(i))$, transactions related to distribution of CC and updated reputation of vehicles. As shown in Fig. 5, the blocks are committed by $RLY(i)$ at each hop. Every vehicle in permissioned network possesses the blockchain. Addition of block is announced by $RLY(i)$ to vehicles in its transmission range. Each vehicle is responsible for updating its blockchain and coordinating it with CA regularly.

CC is used as a monetary incentive in the proposed approach. It is divided into two parts with ratio $w1 : w2$, where w_1 and w_2 are corresponding weights to divide the share of CC among $\overline{END}(0)$ and $RLY(i)$, $RLY(i+1)$,…,$RLY(N_{HOP}^{max})$ respectively and $w_1 + w_2 = 1$. The profit, P_j of a vehicle j is given as

$$P_j = \begin{cases} \frac{w_1 CC}{N_{END}^{min}}, & \text{if } j \epsilon \overline{END}(0), \\ \frac{w_2 CC}{N_{HOP}}, & \text{if } j = RLY(i), \\ 0, & \text{otherwise.} \end{cases} \tag{9}$$

The transactions into credit wallets of $\overline{END}(i)$ are committed as a block by $RLY(i)$ only when $i = 0$ and transactions into credit wallets of $RLY(i)$, $RLY(i+1)$,…, $RLY(N_{HOP}^{max})$ are committed as a block at last hop by $RLY(N_{HOP}^{max} + 1)$. If N_{END}^{min} *endorsements* are not obtained for a *transaction proposal* until t_{END}^{max}, it is considered as fake and CC is transferred to CA as a penalty to ORG. The penalty

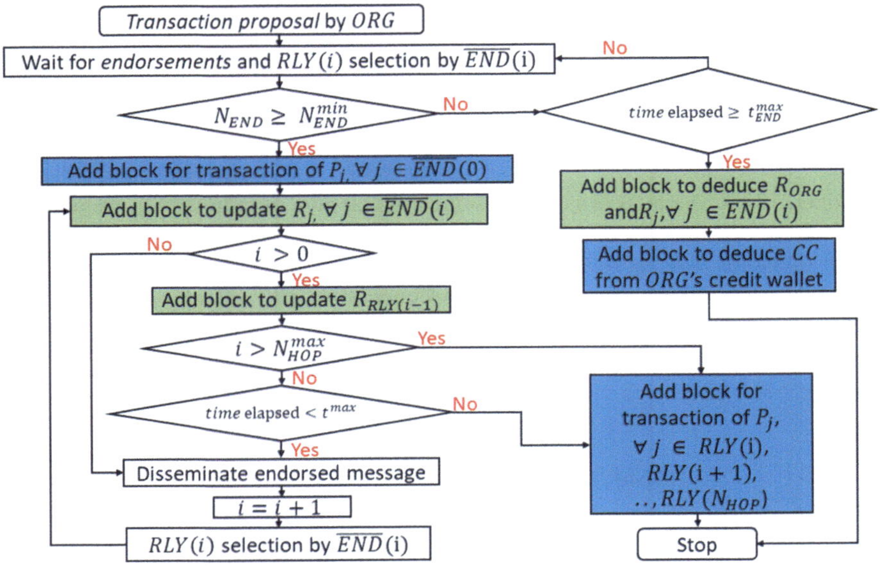

Fig. 5 Flowchart of blockchain-enabled message dissemination in IoV

is recorded as a transaction in the blockchain by $RLY(i)$ which is selected by ORG itself while originating a *transaction proposal*. No share of CC is paid to $\overline{END}(0)$ which endorsed a fake *transaction proposal*.

Reputation is affected by the behaviour of vehicle. Honest behaviour of $\overline{END}(i)$ and $RLY(i)$ is recognised as the successful action performed during message dissemination. Malicious behaviour refers to a fake *transaction proposal* initiated by ORG or endorsed by $\overline{END}(i)$ and selfish $RLY(i)$ without disseminating an endorsed message. Reputation updates are committed as transactions in the blockchain by $RLY(i)$ at each hop i. $R_{\overline{END}(i)}$ are updated at ith hop and $R_{RLY(i)}$ is updated at $(i + 1)$th hop. If $RLY(i)$ itself behaves maliciously, its reputation deduction is processed in the blockchain by CA after being reported by $\overline{END}(i)$. If a vehicle j behaves honestly, its reputation is updated with

$$R_j = R_j + \beta, \tag{10}$$

where β is the reputation reward. If it behaves maliciously, its reputation is updated with

$$R_j = R_j - \gamma, \tag{11}$$

where γ is the reputation penalty. The credit and reputation management follows an economic model defined in Table 3.

Table 3 Economic model for credit and reputation management

Vehicle	Credit		Reputation	
	Honest	Malicious	Honest	Malicious
ORG	$-TC - CC$	$-TC - CC$	–	$R_{ORG} - \gamma$
$j \epsilon \overline{END}(i = 0)$	$P_j - TC$	$-TC$	$R_{\overline{END}(i)} + \beta$	$R_{\overline{END}(i)} - \gamma$
$j \epsilon \overline{END}(i > 0)$	–	–	$R_{\overline{END}(i)} + \beta$	–
$j = RLY(i)$	$P_j - TC$	0	$R_{RLY(i)} + \beta$	$R_{RLY(i)} - \gamma$

Game Theory Analysis

We apply game theory to analyse the performance of our incentive distribution mechanism against collusion of $RLY(i)$, $RLY(i + 1)$,...,$RLY(n)$, where n is equivalent to the number of colluding relay nodes. We find the best action for players and associated conditions so as to provide positive utility to honest players. By setting up a suitable condition, we can guarantee the security of our solution against colluding behaviour of relay nodes. The model of our blockchain-enabled message dissemination game is described as follows

- **Players** This game has N_{END}^{min} number of $\overline{END}(0)$ and one $RLY(i)$ at each hop i. The number of hops is N_{HOP}.
- **Actions** At hop $i = 0$, $j \epsilon \overline{END}(0)$ has three possible actions, honest (H), malicious (M) and selfish (S). If it votes for a true message, it is honest. If it votes for a false message, it is malicious. If it does not cooperate, it is selfish. At each hop i, $RLY(i)$ has two possible actions: honest (H) and selfish (S). If it forwards the message and commits block, it is honest. If it does not cooperate, it is selfish. We denote the action of player j by ACT_j, which is either H, M or S.
- **Utilities** Without colluding with its neighbours, the utility U_j is

$$U_j = \begin{cases} -TC, & \text{if } j \epsilon \overline{END}(0), ACT_j = M, \\ P_j - TC, & ACT_j = H, \\ 0, & \text{otherwise.} \end{cases} \tag{12}$$

We present the following definitions for the security analysis of our incentive distribution mechanism.

Definition 1 The best response action for a player is such that it brings the maximum expected utility to itself, regardless of the actions of all other players.

Definition 2 The incentive distribution mechanism is $RLY(i)$ collusion resistant if $RLY(i)$ and any group of its colluding neighbours cannot increase the expected sum of their utilities by using any action other than the one in which everybody plays honestly.

The following proposition and theorem are also presented here.

Proposition 1 *In blockchain-enabled message dissemination game, playing honest is the best response action of all players if* $0 \leq TC \leq P_j$.

Proof According to (12), if $TC < 0$, the utility of player j, for $j \epsilon \overline{END}(i)$ would be positive if it plays maliciously. On the other hand, if $TC \geq P_j$, the utility of player j would be negative if it plays honestly. In either of these cases, the best response action of players would be to play maliciously or selfishly. Therefore TC must be set such that the action results in positive U_j or profit gain in credit wallets of all players only if they play honestly. □

Theorem 1 *The incentive distribution mechanism is resistant against collusion of* $RLY(i)$, $RLY(i+1)$,..., $RLY(n)$ *if* $TC \geq 0$.

Proof Lets consider the case with one conspired $RLY(i)$. Suppose $CG = \{RLY(i), RLY(i+1)\}$ is a collusion group. CG conspires to form a bogus path with an additional hop, i.e., $ORG \rightarrow RLY(i) \rightarrow RLY(i+1)$ instead of the most appropriate path, i.e. $ORG \rightarrow RLY'(i+1)$. Let p be the probability with which $RLY(i)$ encounters, where $p \epsilon [0, 1]$. Therefore, p^2 is the probability with which it encounters both ORG and $RLY(i+1)$. The expected utility sum of CG, $E(U_{CG})$ is

$$E(U_{CG}) = p^2(U_{RLY(i)} + U_{RLY(i+1)}) + (1 - p^2)(U_{RLY'(i+1)}), \tag{13}$$

or,

$$\begin{aligned} E(U_{CG}) = p^2(&P_{RLY(i)} - TC + P_{RLY(i+1)} - TC) \\ &+ (1 - p^2)(P_{RLY'(i+1)} - TC). \end{aligned} \tag{14}$$

Since $N_{HOP} = 2$ for collusion case and $N_{HOP'} = 1$ for non-collusion, $E(U_{CG}$ becomes

$$E(U_{CG}) = p^2\Big(\frac{w_2CC}{2} + \frac{w_2CC}{2} - 2TC\Big) + (1 - p^2)(w_2CC - TC), \tag{15}$$

or,

$$E(U_{CG}) = w_2CC - TC - p^2TC. \tag{16}$$

To avoid collusion of relay nodes, we want $E(U_{CG}) \leq U_{RLY'(i+1)}$, that is,

$$w_2CC - TC - p^2TC \leq w_2CC - TC. \tag{17}$$

It follows that

$$-p^2TC \leq 0, \tag{18}$$

or,

$$p^2TC \geq 0. \quad (19)$$

Similarly, by generalising cases when $n > 2$, we can derive the collusion resistant condition, that is, $p^nTC \geq 0$. Hence, for any $p\epsilon[0, 1]$, we can prove that the incentive distribution mechanism is relay node collusion resistant if $TC \geq 0$. □

4.4 Annual Road Tax

The blockchain-enabled IoV can also be used to calculate road tax. It can be annually calculated on the basis of remaining balance of each vehicle's credit wallet at the end of the year. The motivation behind this approach is due to the amount in a credit wallet reflecting a vehicle's behaviour. A vehicle which is involved in less number of incidents would have spent less credit in originating $transactionproposals$, leaving higher balance remaining in its credit wallet which can be redeemed into road tax. Furthermore, the vehicles which are near to the incident's location would be motivated to take part in message dissemination and earn the credit as a compensation of being affected by the incident.

5 Simulation Results and Discussion

In this section, the performance of our blockchain-enabled solution is discussed on the basis of results obtained through extensive simulations using OMNeT++ integrated with SUMO (Simulation of Urban Mobility) which can be seen in Fig. 6.

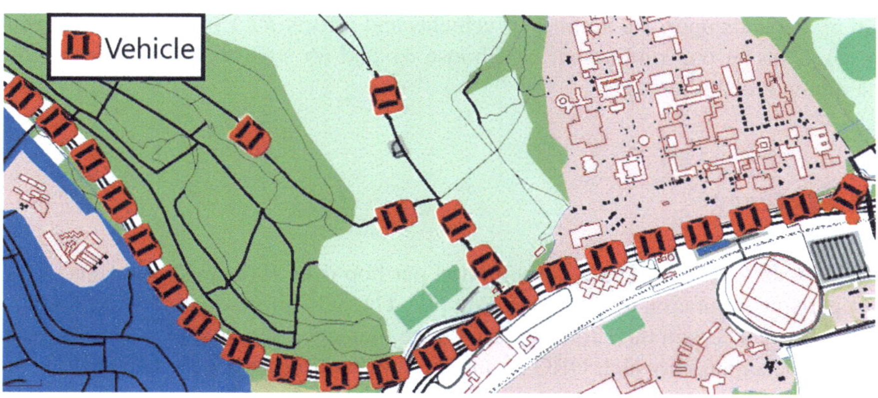

Fig. 6 Simulation map of University of Sussex campus

Table 4 Simulation parameters

Parameter	Value	Parameter	Value
Run time	1000 s	Size of area	12.5× 12.5 km^2
No. of vehicles	[50, 200]	Cryptography	SHA-256
Protocol	IEEE 802.11p	Average speed	40 km/h
Data rate	6 Mbps	Sensitivity	−89 dBm
t^{max}_{END}	600 ms	N^{min}_{END}	3
N^{max}_{HOP}	6	CC	$\frac{10}{R_{ORG}}$
w_1	0.35	w_2	0.65
TC	1	R_j	[0,1]
R_T	0.5	β	0.1
γ	0.2	α_1	[1, 10]
α_2	[1, 4]	d^{min}_{HOP}	100 m
t_W	10 s	t_{max}	4 s

5.1 Simulation Setup

The simulation parameters in Table 4 are set so that Proposition 1 holds true regardless of the value of CC which is inversely proportional to R_{ORG}. In our simulations, we have arbitrarily set $CC = 10/R_{ORG}$. Higher R_{ORG} leads to lower amount of CC, thereby resulting in less profit for $\overline{END}(0)$ and $RLY(i)$, $RLY(i+1)$,...,$RLY(N^{max}_{HOP})$. However, it must be ensured that the message dissemination solution results in profit gain in credit wallets of all, despite of deduction of TC. As shown in Fig. 7, $w_1 > 0.3$ and $w_2 > 0.6$ would always result in positive profit, P_j, irrespective of the value R_{ORG}. Therefore, in order to ensure profit gain, we have set $w_1 = 0.35$ and $w_2 = 0.65$ in simulation. For efficient $RLY(i)$ selection, α_1 and α_2 have to be optimised. Figure 8 shows the percentage of vehicles which received the message within a specified period of time, t_{max}, with respect to α_1 and α_2 under different traffic densities. It shows that the selection of α_1 and α_2 depends upon the number of vehicles and affects $RLY(i)$ selection defined in (2). Thus in our simulation, we choose α_1 and α_2 such that they achieve the maximum reception rate.

5.2 Latency

Figure 9 shows average time consumption per hop over 100 simulation runs. As a comparison, reputation based blockchain [37, 41], only allowed vehicles with reputation above a certain threshold to disseminate messages. CreditCoin was witnesses based blockchain, in which ORG required a threshold number of witnesses to vote for authentication of a transaction proposal [34] and then allowed only authentic transaction proposals to be forwarded as messages. The latency was increased in

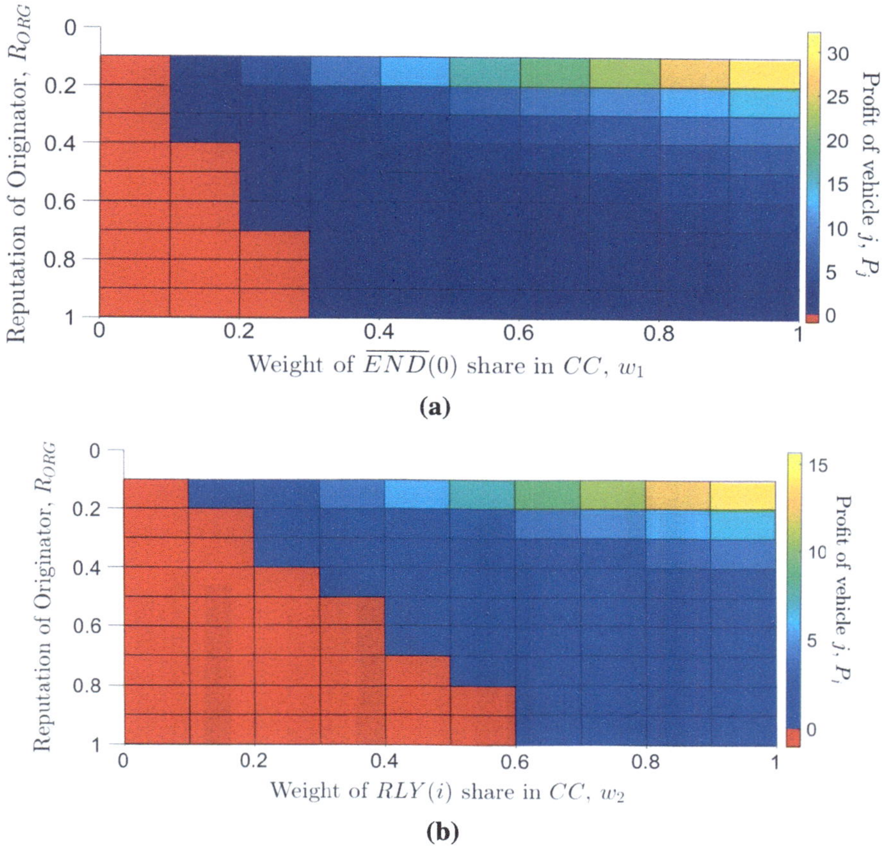

Fig. 7 P_j with respect to R_{ORG}, w_1 and w_2. (**a**) P_j for $j \epsilon \overline{END}(i)$. (**b**) P_j for $j = RLY(i)$

this approach because it relied on ORG to wait for the witnesses before generating the actual announcement packet. On the contrary, in our approach, the integrated authentication and relaying is implemented in a distributed manner and therefore, waiting time of ORG to rebroadcast after authentication is eliminated, which saves an average of 56 ms (or 11%) of time consumed in each consensus. Moreover, in CreditCoin, transactions in blockchain were processed by RSU, whereas in our solution, this task is performed by $RLY(i)$. Figure 9 also shows that reputation based method takes the least time to complete one hop as it does not involve waiting time for endorsements. The only time it consumes is to access blockchain to find reputation of ORG. However, average time increases with raising number of vehicles. This is because when there is a large number of vehicles registered in a blockchain network, it takes more time to access and find reputation of a vehicle. In proposed approach and CreditCoin, average time decreases with raising number of vehicles. This is because N_{END}^{min} endorsements are gathered in less time in heavy

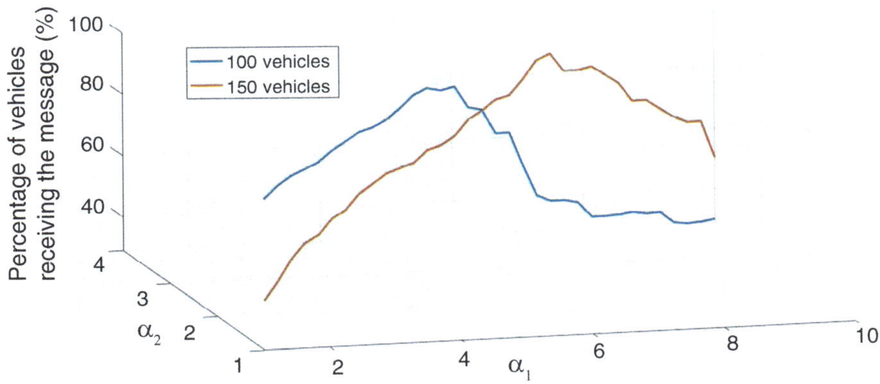

Fig. 8 Percentage of vehicles which received message with respect to α_1 and α_2

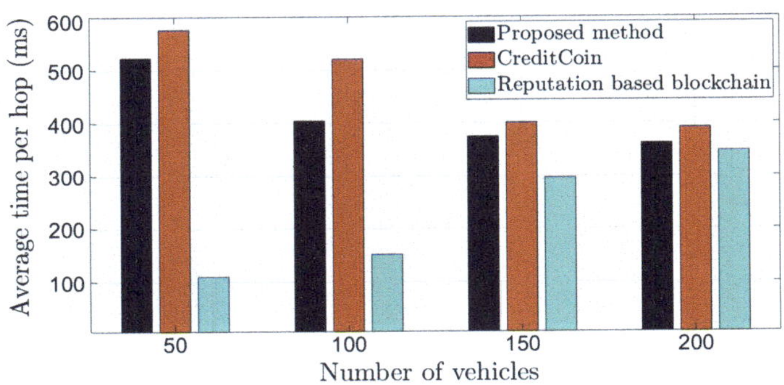

Fig. 9 Average time consumption per hop

traffic conditions. Therefore, time difference between reputation and voting based approaches becomes negligible with large number of vehicles in a network. With 50 vehicles, our solution consumes 522 ms for completing one consensus process. Given the average speed of vehicle is 40 km/h, it only incurs a moving distance of approximately 5 m, which can be easily mitigated within 300 m coverage of typical IEEE 802.11p radio [54]. Therefore, the proposed solution can be practically applied and vehicles are not likely to lose connectivity during a consensus process. The worst case scenario is presented in Fig. 10, where low density of vehicles, i.e. 50 and 100, at speeds beyond 100 and 110 km/h, respectively, cannot successfully complete a consensus algorithm within the time limit of 600 ms, which is practically suitable for threshold based authentication methods in VANETs [55]. However, such high speeds are not likely to be attained in an area affected by an incident or traffic jam. Overall, it shows that the proposed approach is suitable for vehicular networks, particularly for high density traffic with lower speed.

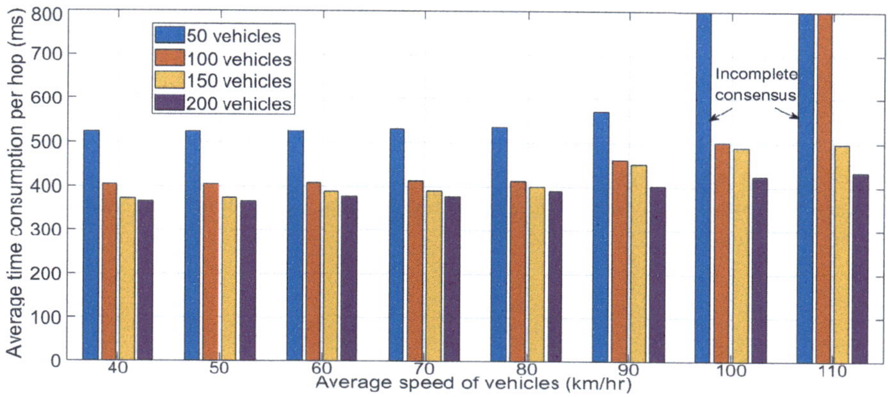

Fig. 10 Average time consumption per hop with respect to speed and number of vehicles

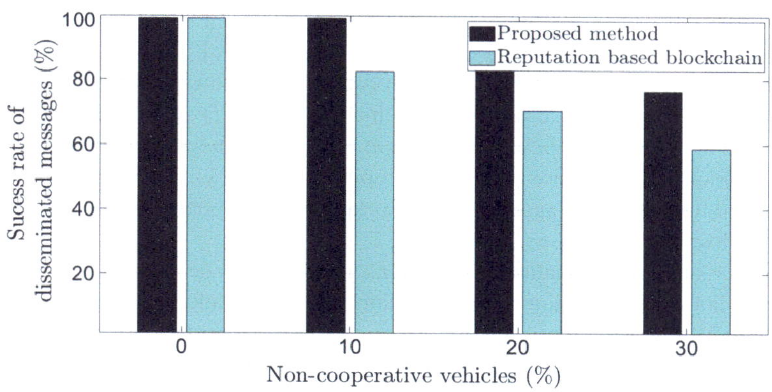

Fig. 11 Success rate of message dissemination in presence of low-reputed vehicles

5.3 *Success Rate*

Figure 11 shows number of messages disseminated successfully in presence of low-reputed vehicles. In our simulation, we have considered low-reputed vehicles as those vehicles whose reputation fall below R_T due to their malicious behaviour. The reputation based blockchain approaches [37, 41] authenticated a message on the basis of reputation of ORG. They did not allow a low-reputed vehicle to originate a transaction proposal in order to prevent dissemination of potentially false messages. This is how they discouraged vehicles with low reputation to contribute actively in blockchain extension and ultimately preventing some valid transaction proposals to be disseminated. Our approach authenticates transaction proposals through voting based consensus algorithm. Therefore, a low-reputed vehicle can also originate a transaction proposal. The trust among vehicles is still maintained because a message cannot be further disseminated without getting N_{END}^{min} endorsements for a transaction proposal. The proposed consensus algorithm results in dissemination of greater number of messages despite the presence of higher percentage of low-reputed vehicles in a network, by equally promoting all vehicles to contribute in blockchain extension even if they have low reputation values. Our proposed approach disseminates on an average 17% more authenticated messages as compared to reputation based blockchain approach in presence of low-reputed vehicles.

5.4 Complexity

To support a completely decentralised message dissemination solution, each vehicle in the reputation based blockchain has the whole copy of blockchain in order to find reputation of any vehicle whenever needed. Therefore, the storage complexity of reputation based blockchain is $O(B)$, where B is the total number of blocks in a blockchain [56]. On the other hand, our proposed solution does not require each vehicle to store the whole copy of blockchain for $RLY(i)$ selection and message dissemination. To add a block in blockchain, the vehicle needs only the address of previous block and therefore its storage complexity is $O(1)$. However, all vehicles are required to update and synchronise their last block responsibly in order to avoid forks and discrepancies.

In terms of communication, the conventional PBFT requires at least N_{END}^{min} signatures both during endorsement and committing a block, as shown in Fig. 12a. Therefore, PBFT results in an overall communication complexity of $O((N_{END}^{min})^2)$ [57], whereas our proposed solution, shown in Fig. 12b, requires N_{END}^{min} signatures only during endorsement and hence results in communication complexity of $O(N_{END}^{min})$.

6 Conclusion

This chapter presents the concept of implementing blockchain technology for secure and private communications in IoV. Various blockchain types and their consensus algorithms are compared on the basis of their suitability in vehicular applications. The advantages of using blockchain for enhancing security and privacy of IoT and IoV are also discussed.

This chapter further proposes a voting based consensus algorithm incentivising vehicles through credit and reputation rewards for message dissemination in emergency situations. For efficient message dissemination, relay selection is made a part of consensus. The proposed solution is analysed by game theory and is proved to be secure against collusion of relay nodes. Further evaluation is conducted through simulations and results show that it saves on an average of 11% time consumption

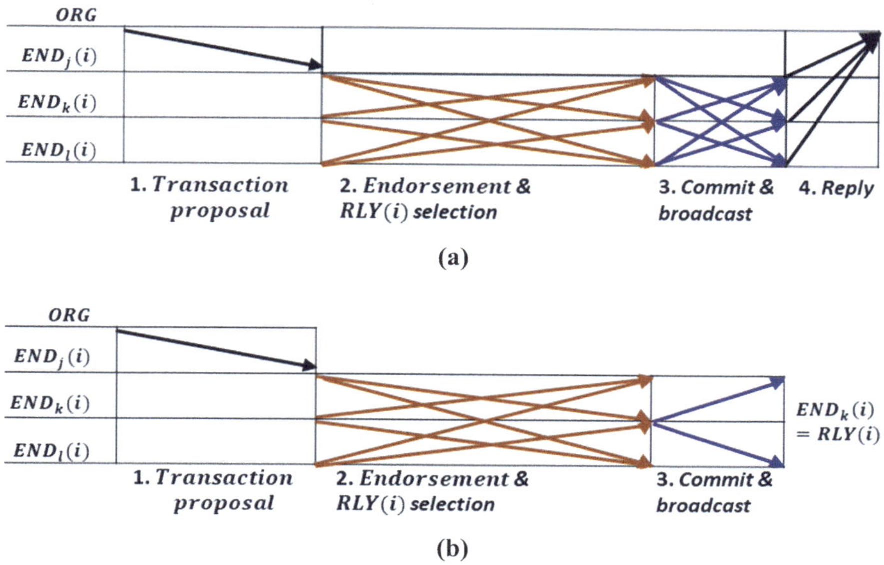

Fig. 12 Flow of voting based consensus algorithms. (**a**) Conventional PBFT. (**b**) Proposed

in authenticating and disseminating message as compared to another voting based validation method. Moreover, it improves successful dissemination of authenticated messages by 17% as compared to reputation based blockchain. As a trade-off, it requires more time to generate block. However, the latency difference is negligible with increasing number of vehicles. The complexity of proposed solution is also less than reputation based consensus and conventional PBFT based consensus algorithm in terms of storage and communication respectively. However, PBFT is less secure as compared to PoW. The design of future blockchain models for IoV should include improved security, reduced latency and complete independence from central authorities.

References

1. Aldegheishem, A., Yasmeen, H., Maryam, H., Shah, M. A., Mehmood, A., Alrajeh, N., et al. (2018). Smart road traffic accidents reduction strategy based on intelligent transportation systems (TARS). *Sensors, 18*(7), 1983–2006.
2. Chen, R., Jin, W. L., & Regan, A. (2010). Broadcasting safety information in vehicular networks: Issues and approaches. *IEEE Network, 24*(1), 20–25.
3. Yáñez, A., Céspedes, S., & Rubio-Loyola, J. (2018). CaSSaM: Context-aware system for safety messages dissemination in VANETs. In *Abstracts of the IEEE Colombian Conference on Communications and Computing, Medellin, Colombia, 16–18 May 2018.*

4. Martuscelli, G., Boukerche, A., Foschini, L., & Bellavista, P. (2016). V2V protocols for traffic congestion discovery along routes of interest in VANETs: A quantitative study. *Wireless Communications and Mobile Computing, 16*(17), 2907–2923.
5. Cataldi, P., & Harri, J. (2011). User/operator utility-based infrastructure deployment strategies for vehicular networks. In *Abstracts of the IEEE Vehicular Technology Conference, San Francisco, CA, USA, 5–8 September 2011*.
6. Yasser, A., Elzorkany, M., & Kader, N. A. (2017). Vehicle to vehicle implementation in developing countries. In A. Hassanien, K. Shaalan, T. Gaber, & M. Tolba (Eds.), *Proceedings of the International Conference on Advanced Intelligent Systems and Informatics 2017. AISI 2017. Advances in Intelligent Systems and Computing* (Vol. 639, pp. 809–819). Cham: Springer.
7. Kazmi, A., Khan, M. A., & Akram, M. U. (2016). DeVANET: Decentralized software-defined VANET architecture. In *Abstracts of the IEEE International Conference on Cloud Engineering Workshop, Berlin, Germany, 4–8 April 2016*.
8. Gazdar, T., Belghith, A., & Abutair, H. (2017). An enhanced distributed trust computing protocol for VANETs. *IEEE Access, 6*, 380–392.
9. Zhang, L., Wu, Q., Domingo-Ferrer, J., Qin, B., & Hu, C. (2016). Distributed aggregate privacy-preserving authentication in VANETs. *IEEE Transactions on Intelligent Transportation Systems, 18*(3), 516–525.
10. Manvi, S. S., & Tangade, S. (2017). A survey on authentication schemes in VANETs for secured communication. *Vehicular Communications, 9*, 19–30.
11. Rehman, O., & Ould-Khaoua, M. (2019). A hybrid relay node selection scheme for message dissemination in VANETs. *Future Generation Computer Systems, 93*, 1–17.
12. He, Y., Yu, F. R., Wei, Z., & Leung, V. (2019). Trust management for secure cognitive radio vehicular ad hoc networks. *Ad-hoc Networks, 86*, 154–165.
13. Hasrouny, H., Samhat, A. E., Bassil, C., & Laouiti, A. (2019). Misbehavior detection and efficient revocation within VANET. *Journal of Information Security and Applications, 46*, 193–209.
14. Janzadeh, H., Fayazbakhsh, K., & Dehghan, M. (2008). A secure credit-based cooperation stimulating mechanism for MANETs using hash chains. *Future Generation Computer Systems, 25*(8), 926–934.
15. Crowcroft, J., Gibbens, R., Kelly, F., & Östring, S. (2004). Modelling incentives for collaboration in mobile ad hoc networks. *Performance Evaluation, 57*(4), 427–439.
16. Refaei, M. T., DaSilva, L. A., Eltoweissy, M., & Nadeem, T. (2010) Adaptation of reputation management systems to dynamic network conditions in ad hoc networks. *IEEE Transactions on Computers, 59*(5), 707–719.
17. Dewan, P., Dasgupta, P., & Bhattacharya, A. (2004). On using reputations in ad hoc networks to counter malicious nodes. In *Abstracts of the IEEE 10th International Conference on Parallel and Distributed Systems, Newport Beach, CA, USA, 9 July 2004*.
18. Srivastava, V., Neel, J., MacKenzie, A. B., Menon, R., DaSilva, L. A., Hicks, J. E., et al. (2005). Using game theory to analyze wireless ad hoc networks. *IEEE Communications Surveys and Tutorials, 7*(4), 46–56.
19. Li, Z., & Shen, H. (2011). Game-theoretic analysis of cooperation incentive strategies in mobile ad hoc networks. *IEEE Transactions on Mobile Computing, 11*(8), 1287–1303.
20. Nakamoto, S. (2008). Bitcoin: A peer-to-peer electronic cash system. Retrieved February 09, 2020, from https://bitcoin.org/bitcoin.pdf
21. Jindal, A., Aujla, G. S., & Kumar, N. (2019). SURVIVOR: A blockchain based edge-as-a-service framework for secure energy trading in SDN-enabled vehicle-to-grid environment. *Computer Networks, 153*, 36–48.
22. Wang, X., Zha, X., Ni, W., Liu, R. P., Guo, Y. J., Niu, X., et al. (2019). Survey on blockchain for Internet of Things. *Computer Communications, 136*, 10–29.
23. Zheng, Z., Xie, S., Dai, H., Chen, X., & Wang, H. (2017). An overview of blockchain technology: Architecture, consensus, and future trends. In *Abstracts of the IEEE International Congress on Big Data, Honolulu, HI, USA, 25–30 June 2017*.

24. Gao, W., Hatcher, W. G., & Yu, W. (2018). A survey of blockchain: Techniques, applications, and challenges. In *Abstracts of the IEEE International Conference on Computer Communication and Networks, Hangzhou, China, 30 July–2 August 2018*.
25. Decker, C., & Wattenhofer, R. (2013). Information propagation in the bitcoin network. In *Abstracts of the IEEE International Conference on Peer-to-Peer Computing, Trento, Italy, 9–11 September 2013*.
26. Ortega, V., Bouchmal, F., & Monserrat, J. F. (2018). Trusted 5G vehicular networks: Blockchains and content-centric networking. *IEEE Vehicular Technology Magazine, 13*(12), 121–127.
27. Sankar, L. S., Sindhu, M., & Sethumadhavan, M. (2017). Survey of consensus protocols on blockchain applications. In *Abstracts of the IEEE 4th International Conference on Advanced Computing and Communication Systems (ICACCS), Coimbatore, India, 6–7 January 2017*.
28. Fairley, P. (2018). Ethereum will cut back its absurd energy use. *IEEE Spectrum, 56*(1), 29–32.
29. Baliga, A. (2017). Understanding blockchain consensus models. *Persistent, 4*, 1.
30. Wang, W., Hoang, D. T., & Hu, P. (2019). A survey on consensus mechanisms and mining strategy management in blockchain networks. *IEEE Access, 7*, 22328–22370.
31. Muratov, F., Lebedev, A., Iushkevich, N., Nasrulin, B., & Takemiya, M. (2018). YAC: BFT consensus algorithm for blockchain. Preprint. arXiv:1809.00554.
32. Dagher, G. G., Mohler, J., Milojkovic, M. (2018). Ancile: Privacy-preserving framework for access control and interoperability of electronic health records using blockchain technology. *Sustainable Cities and Society, 39*, 283–297.
33. Samuel, O., Javaid, N., Shehzad, F., Iftikhar, M. S., Iftikhar, M. Z., Farooq, H., et al. (2020). Electric vehicles privacy preserving using blockchain in smart community. In L. Barolli, P. Hellinckx, & T. Enokido (Eds.), *Advances on Broad-band Wireless Computing, Communication and Applications. BWCCA 2019. Lecture Notes in Networks and Systems* (Vol. 97). Cham: Springer.
34. Li, L., Liu, J., Cheng, L., Qiu, S., Wang, W., Zhang, X., et al. (2018). Creditcoin: A privacy-preserving blockchain-based incentive announcement network for communications of smart vehicles. *IEEE Transactions on Intelligent Transportation Systems, 19*(7), 2204–220.
35. Lu, Z., Wang, Q., Qu, G., et al. (2018). Bars: a blockchain-based anonymous reputation system for trust management in VANETs. In *Abstracts of the IEEE 17th International Conference on Trust, Security and Privacy in Computing and Communications/12th International Conference on Big Data Science and Engineering, New York, NY, USA, 1–3 August 2018*.
36. Liu, L., & Loper, M. (2018). Trust as a service: building and managing trust in the internet of things. In *Abstracts of the IEEE International Symposium on Technologies for Homeland Security (HST), Woburn, MA, USA, 23–24 October 2018*.
37. Kang, J., Xiong, Z., Niyato, D., Ye, D., Kim, D. I., & Zhao, J. (2018). Toward secure blockchain-enabled internet of vehicles: Optimizing consensus management using reputation and contract theory. *IEEE Transactions on Vehicular Technology, 68*(3), 2906–2920.
38. Yang, Z., Yang, K., Lei, L., Zheng, K., & Leung, V. C. (2018). Blockchain-based decentralized trust management in vehicular networks. *IEEE Internet of Things Journal, 6*(2), 1495–1505.
39. Khan, A. S., Balan, K., Javed, Y., Tarmizi, S., & Abdullah, J. (2019) Secure trust-based blockchain architecture to prevent attacks in VANET. *Sensors, 19*(22), 4954.
40. Ali, N. A., Taha, A. E. M., & Barka, E. (2020). Integrating blockchain and IoT/ITS for safer roads. *IEEE Network, 34*(1), 32–37.
41. Yang, Z., Zheng, K., Yang, K., & Leung, V. C. (2017). A blockchain-based reputation system for data credibility assessment in vehicular networks. In *Abstracts of the IEEE 28th Annual International Symposium on Personal, Indoor, and Mobile Radio Communications, Montreal, QC, Canada, 8–13 October 2017*.
42. Tian, F. (2017). A supply chain traceability system for food safety based on HACCP, blockchain & Internet of things. In *Abstracts of the International Conference on Service Systems and Service Management, Dalian, China, 16–18 June 2017*.
43. Sharma, P. K., Moon, Y. S., & Park, H. J. (2017). Block-VN: A distributed blockchain based vehicular network architecture in smart city. *Journal of Information Processing Systems, 13*(1), 184–195.

44. Dao, T. C., Nguyen, B. M., & Do, B. L. (2019). Challenges and strategies for developing decentralized applications based on blockchain technology. In L. Barolli, M. Takizawa, F. Xhafa, & T. Enokido (Eds.), *Advanced Information Networking and Applications. AINA 2019. Advances in Intelligent Systems and Computing* (Vol. 926). Cham: Springer.
45. Leiding, B., Memarmoshrefi, P., & Hogrefe, D. (2016). Self-managed and blockchain-based vehicular ad-hoc networks. In *Abstracts of the ACM International Joint Conference on Pervasive and Ubiquitous Computing: Adjunct, Heidelberg, Germany, 12–16 September 2016.*
46. Alouache, L., Nguyen, N., Aliouat, M., & Chelouah, R. (2018). Credit based incentive approach for V2V cooperation in vehicular cloud computing. In A. Skulimowski, Z. Sheng, S. Khemiri-Kallel, C. Cérin, & C. H. Hsu (Eds.), *Internet of Vehicles, Technologies and Services Towards Smart City. IOV 2018. Lecture Notes in Computer Science* (Vol. 11253, pp. 92–105). Cham: Springer.
47. Ayaz, F., Sheng, Z., Tian, D., Guan, YL., & Leung, V. (2020, accepted). A voting blockchain based message dissemination in vehicular ad-hoc networks (VANETs). In *IEEE International Conference on Communications, Virtual Conference, 7–11 June 2020.*
48. Pescaru, D. (2013). Urban traffic congestion prediction based on routes information. In *Abstracts of the IEEE 8th International Symposium on Applied Computational Intelligence and Informatics (SACI), Timisoara, Romania, 23–25 May 2013.*
49. Abedi, O., Berangi, R., Azgomi, M. A. (2009). Improving route stability and overhead on AODV routing protocol and make it usable for VANET. In *Abstracts of the 29th International Conference on Distributed Computing Systems Workshops, Montreal, QC, Canada, 22–26 June 2009.*
50. Iza-Paredes, C., Mezher, A. M., Aguilar Igartua, M., & Forné, J. (2018). Game-theoretical design of an adaptive distributed dissemination protocol for VANETs. *Sensors, 18*(1), 294–325.
51. Fatemidokht, H., & Rafsanjani, M. K. (2018). F-Ant: An effective routing protocol for ant colony optimization based on fuzzy logic in vehicular ad hoc networks. *Neural Computing and Applications, 29*(11), 1127–1137.
52. Bettstetter, C., Hartenstein, H., & Pérez-Costa, X. (2004). Stochastic properties of the random waypoint mobility model. *Wireless Networks, 10*(5), 555–567.
53. Chatterjee, S., & Das, S. (2015). Ant colony optimization based enhanced dynamic source routing algorithm for mobile ad-hoc network. *Information Science, 295*, 67–90.
54. Boban, M., & d'Orey, P. M. (2016). Exploring the practical limits of cooperative awareness in vehicular communications. *IEEE Transactions on Vehicular Technology, 65*(6), 3904–3916.
55. Shao, J., Lin, X., Lu, R., & Zuo, C. (2015). A threshold anonymous authentication protocol for VANETs. *IEEE Transactions on Vehicular Technology, 65*(3), 1711–1720.
56. Ahmad, A., Saad, M., Njilla, L., Kamhoua, C., Bassiouni, M., & Mohaisen, A. (2018). Blocktrail: A scalable multichain solution for blockchain-based audit trails. In *Abstracts of the IEEE International Conference on Communications (ICC), Shanghai, China, 20–24 May 2019.*
57. Choi, B., Sohn, J. Y., Han, D. J., & Moon, J. (2019). Scalable network-coded PBFT consensus algorithm. In *Abstracts of the IEEE International Symposium on Information Theory (ISIT), Paris, France, 7–12 July 2019.*

Approximation Algorithm and Linear Congruence: A State-of-Art Approach in Information Security Issues Towards Internet of Vehicles

Anirban Bhowmik, Sunil Karforma, Joydeep Dey, and Arindam Sarkar

1 Introduction

Vehicle communications are becoming increasingly popular issue. Smart vehicular application or IoV has been a part of intelligent transportation system (ITS) [1, 2], and its aim is to provide a safer, coordinated, and smarter mode of transportation. With the help of smart vehicular system, ITS can improve traffic management efficiently and enhance road safety. Smart vehicular system uses a vehicle-to-vehicle (V2V), vehicle-to-infrastructure (V2I) communication, and VANET [3] to obtain traffic and vehicle status information in advance which helps to reduce the traffic accidents in road. Multi-hop transmission [4] technique is used to exchange information among vehicles. V2V communication is based on dedicated short-range communication (DSRC) [5] standard. In this context smart vehicular application is an open and different application based system with insecure communication environment. The main goal of smart vehicular application is to improve traffic safety by which we can transmit messages among different vehicles efficiently and honestly. Smart vehicular system is vulnerable to various attacks. Therefore, in

A. Bhowmik (✉)
Department of Computer Application, Cyber Research and Training Institute, Burdwan, West Bengal, India

S. Karforma
Department of Computer Science, The University of Burdwan, Burdwan, West Bengal, India

J. Dey
Department of Computer Science, M.U.C. Women's College, Burdwan, West Bengal, India

A. Sarkar
Department of Computer Science and Electronics, R.K.M. Vidyamandira, Belur Math, Belur, West Bengal, India

© Springer Nature Switzerland AG 2021

N. Gupta et al. (eds.), *Internet of Vehicles and its Applications in Autonomous Driving*, Unmanned System Technologies, https://doi.org/10.1007/978-3-030-46335-9_10

V2V communication, verification, authentication, and security of the messages in communication network are very necessary.

Cryptography and Secret Keys Cryptography [6] deals with creating documents that can be shared secretly over public communication channels. It is study of creating and using encryption and decryption techniques. An encryption algorithm works with a key to transform the plain text into cipher text so that hackers cannot read it, but that can be authorized person. The decryption algorithm works in the reverse order and converts the cipher text into plain text. Now key is the number which is used to create cipher text from plain text. Cryptography is grouped into two parts such as symmetric key and asymmetric key [6, 7] cryptography.

RSA The RSA [6, 8] algorithm is a suite of cryptographic techniques that are used for specific security purposes which enables public key encryption. It is widely used to secure sensitive data, particularly when it is sent over an insecure network such as Internet. RSA was first publicly described in 1977 by Ron Rivest, Adi Shamir, and Leonard Adleman of the Massachusetts Institute of Technology. In RSA cryptography, both the public and the private keys can encrypt a message the opposite key from the one used to encrypt a message is used to decrypt it. RSA has become the most widely used asymmetric algorithm because it provides a method to assure the confidentiality, integrity, authenticity, and non-repudiation of electronic communications and data storage.

Approximation Algorithm and Subset Sum Problem An algorithm that returns near optimal solutions is called an approximation algorithm [9]. Depending on problem, an optimal solution may be defined as one with maximum possible cost or one with minimum possible cost, i.e., the problem may be either a maximization or minimization problem. An algorithm for a problem has an approximation ratio of $p\ (n)$ if, for any input of size n, the cost C of the solution produced by the algorithm is within a factor of $p\ (n)$ of the cost C_* of an optimal solution: $max(C/C_*, C * /C) \leq p\ (n)$. An algorithm that achieves an approximation ratio $p\ (n)$ is a $p\ (n)$-approximation algorithm.

Definition: An instance of the subset sum decision problem is (S, t) where

$S = \{x_1, x_2, \ldots, x_n\}$ Set of positive integers and t a positive integer.

The problem is whether some subset of S adds up exactly to t. This problem is NP-complete.

The subset sum optimization problem is to find a subset of S whose sum is as large as possible but no greater than t.

We will define a class of algorithms, $A_\in$ such that, $\forall$€ > 0,

- $A_\in$ is an €-approximation algorithm for subset sum.
- $A_\in$ runs in time polynomial in n, log t and 1/€.

Such a class of algorithms is known as a fully polynomial-time approximation scheme.

An exponential time algorithm

If $S = \{x_1, x_2, \ldots, x_n\}$ is a set or list and x a real number then define

$S + x = \{x_1, x_2 \ldots x_n\} = \{x_1 + x, x_2 + x, \ldots, x_n + x\}$.

If $L = \{x_1, x_2, \ldots, x_n\}$ and $L' = \{u_1, u_2, \ldots, u_m\}$ are both sorted lists, then define Merge-Lists (L, L') to be the procedure that returns the sorted union of the two lists. This procedure runs in time$O(|L'| + |L|)$.

Algorithm: (Exact-Subset Sums)

```
Step1. n ← |S|
Step2. L0 ←< 0 >
Step3. for i = 1 to n
Step4. Li = Merge-Lists (Li−1, Li−1 + xi)
Step5. Remove from Li all elements bigger than t.
Step6. Return largest element in Ln.
Step7. Stop
```

Significance of the Subset Sum Problem in Our Scheme The fundamental definition of subset sum problem [9] is to find subsets of a certain sum. Here we enumerate power set of a set which are categorized based on their sum. We consider the set of natural numbers within a range chosen by user as our problem domain.

A set of first n natural numbers is $X_n = \{1, 2, 3, \ldots n\}$ where n is a positive integer. The set X_n is also known as the universal set. This is our problem domain. The cardinality of the set X_n is n. $|X_n| = n$

A set of all subsets of $X_n$is $P(X_n) = \{\emptyset, \{1\}, \{2\}, \ldots, \{1, 2, \ldots n\}\}$. it is also known as power set. The empty set is denoted as Ø or {} or the null set.

Sum Distribution: In sum distribution (SD), we find the number of subsets which sum up to a certain integer S, where $X_n = \{1, 2, 3 \ldots n\}$ and S € $[0, \frac{n(n+1)}{2}]$. It is represented as SD[n][S].

Before counting the subsets of a particular sum, we initialize the count as zero, ¥n, S SD[n][S] = 0.

Following are the base cases for sum distribution (SD[n][S]):

1. For $n = 0$ and $S = 0$, the corresponding subset is Ø. Since, zero-sum (Sum = 0) can be achieved only with subset Ø and Sum (Ø) is assumed to be 0, the count of occurrence of subset in $P(X_0)$ is taken as 1. Therefore, SD [0][0] = 1.
2. ¥i € [1, n] and $S = 0$, SD[i][0] = 1. Since, zero-sum (Sum = 0) can be achieved only with subset Ø, the count of occurrence of Ø-subset in $P(X_n)$ is taken as 1. Therefore, SD[n][0] = 1.
3. SD[i][j] = 0, if i < 0 or j < 0.

   SD[n] [S] =1 when S=0 or n=1.

   SD[n] [S] = SD [n-1] [S] when 0<S<1.

   SD[n] [S] = SD [n-1] [S] + SD [n-1][S-n] when $n < S < \text{FLOOR}(\frac{n(n+1)}{4})$

   SD[n] [S] = SD[n] $[\frac{n(n+1)}{2}$- S]

Linear Congruence Let $f(x) = a_0x^n + a_1x^{n-1} + \ldots + a_n (n \geq 1)$ be a polynomial with integer coefficient a$a_0, a_1, \ldots, a_n$ with $a_0 \not\equiv 0(mod m)$. Then $f(x) \equiv 0(\text{mod } m)$ is said to be a polynomial congruence (mod m) of degree n. If there exists an integer x_0 such that $f(x_0) \equiv 0\ (mod\ m)$, then x_0 is said to be a solution of the congruence [10].

Theorem 1 *If* $gcd\ (a, m) = 1$, *then the linear congruence* $ax \equiv b\ (mod\ m)$ *has a unique solution* [10].

Proof Since $gcd\ (a, m) = 1$, there exist integers u, v such that $au + mv = 1$.

Therefore $a\ (bu) + m\ (bv) = b$. This gives a$(bu) \equiv b\ (mod\ m)$.

This shows that $x = bu$ is a solution of the congruence $ax \equiv b\ (mod\ m)$. Let x_1 and x_2 be solutions of the congruence $ax \equiv b\ (mod\ m)$. Then $ax_1 \equiv b\ (mod\ m)$ and $ax_2 \equiv b\ (mod\ m)$. This implies $ax_1 \equiv ax_2\ (mod\ m) \rightarrow x_1 \equiv x_2\ (mod\ m)$, since $gcd\ (a, m) = 1$. This proves that the congruence has a unique solution.

The concept of the above theorem is used in generation of intermediate key in our proposed technique.

Linear and Non-linear Functions A linear function [10] has the form $y = f(x) = ax + b$. A linear function has one independent variable which is x and one dependent variable which is y. b is constant term or y intercept. a is the coefficient of the independent variable. It is also known as slope and it gives the rate of change of the dependent variable. A linear equation is used in our technique in intermediate key generation.

A simple non-linear equation is of the form $x^2 + by^2 = c$. A non-linear equation [10] looks like a curve when graphed. Its slope value may be changed. The degree of a non-linear equation is at least 2 or other integer values The input and output of a non-linear system are not directly related. In this article a non-linear function is used in encryption process. Here we use non-linear function is to create a non-linearity in cipher text.

Pell's Formula Pell's equation [11] is a Diophantine equation of the form $x^2 - dy^2 = 1$, $x, y \in Z$, where d is a given natural number which is not a square. An equation of the form $x^2 - dy^2 = k$ for an integer k is referred to as a Pell-type equation. It can be solved in the following way. We know that the general solution of a linear Diophantine equation is a linear function of some parameters. Let $d = c^2, c \in Z$, the equation $x^2 - dy^2 = k$ can be factored as $(x - cy)(x + cy) = k$ and therefore it can be solved without using any other theory. So, d will always be assumed not to be a square number and the equation $x^2 - dy^2 = k$ can still be factored as $\left(x + y\sqrt{d}\right)\left(x - y\sqrt{d}\right) = k$. In this case we must deal with the numbers of the form $x + y\sqrt{d}$, where x, y integers. This set is denoted by $Z[\sqrt{d}\,]$. An important property of this set is that the sum and product of two of its elements remain in the set.

2 Literature Survey

In the smart vehicular system the messages are transmitted via wireless medium which is open to attack by intruders. For the protection of transmitted messages following techniques are already introduced. We have discussed them briefly.

1. Y. Hao et al. [3] emphasis on data access privacy protection in vehicles. This article uses bilinear pairings for message encryption and authentication. But it is fact that the bilinear pairing bears higher computational complexity, so, this study uses bilinear pairings for vehicles message authentication or identity verification and applies symmetric-key algorithm for message communication in order to reduce computational complexity for encryption/decryption.
2. Vijayakumar et al. [12] proposed a dual authentication and key management (DAKM) strategy. DAKM scheme uses the concept of dual authentication to prevent unauthenticated messages from unauthorized vehicles. In addition, the DAKM strategy effectively updates messages from vehicles. However, it does not provide the security of the location privacy of vehicles.
3. Chuang and Chen [13] proposed a trust-extended authentication (TEA) technique which authenticates V2V communication. But this technique does not provide any specific way for message authentication and as a result internal attack may appear.
4. Kumari et al. [14] introduced an enhanced TEA (ETEA) strategy which defends against internal attacks but it has a long running time and has heavy computational burden.
5. Li et al. [15] proposed an authentication framework with conditional privacy preservation and non-repudiation (ACPN) in his article. ACPN provides the public key encryption-based pseudonym mechanism which realizes the non-repudiation of vehicles. But, the storage cost of ACPN is high.
6. Wang et al. [16] proposed a technique which focuses on two-factor lightweight privacy-preserving authentication (TFLIP). Here two-factor biological encryption mechanism is used to authenticate the received messages. But, the security of the TFLIP scheme depends heavily on the system secret keys.

3 Problem Domain

From the above literature survey and introduction part we have find that it is easier for attackers to attack the messages in smart vehicular system because here message transmission is done via a wireless medium. If the network has been attacked, transmission delay will be long and the message may have been tampered or lost. A wrong and tampered message in smart vehicular system may cause traffic congestion or even a traffic accident. In addition, the vehicle may receive multiple messages from other vehicles at the same time in a dense environment. Therefore, this creates an urgent problem in the authentication of multiple messages and message security in communication network in a short time. Thus, the message authentication and hacking are a vital issue in the smart vehicular system. Thus we should follow some technique or algorithms so that the system should defend various attacks. Different types of solutions are discussed in the above literature survey but none of which does not fulfill all the security issues. In this paper we have tried to introduce some techniques which are better than previous one.

4 Solution Domain

In this article, we have proposed a solution method to overcome the above problem related to smart vehicular system or IoV. Here we focuses on two basic things related to message communication, i.e., authentication and encryption-decryption. For authentication we have used hash function (MD5) and a key, named, intermediate key which is generated by using the two public keys of sender vehicle and receiver vehicle. For encryption-decryption, two keys (session key and intermediate key) are used which are generated by using the concept of dynamic problem such as subset sum problem and linear congruence. To get the better non-linearity in encryption we have used linear and non-linear functions. The keys are session oriented, i.e., they may vary session to session. A rigorous frame structure is used for message transmission in wireless network. Different types of information are accumulated in this structure and this information is used for identity verification, key generation, and encryption-decryption.

5 Proposed Work

Our model composed of five algorithms which are (a) Intermediate key generation for authentication and encryption, (b) Session key generation for encryption, (c) Encryption, (d) Authentication realization, (e) Decryption process. The summary of our protocol is described below by a compact algorithm.

Algorithm

```
Input: - plaintext.
  Output: - encrypted file.
---------------------------------------------------------
  Methods: -
  Step1. Call SSKG  (  ) // session key generation using
two public keys of two patients.
  Step2. Call INTMKG ( ) // intermediate key generation
for file encryption.
```

```
  Step3. Call ENCYP  (  ) // file encryption process for
generate cipher text for secure transmission.
  Step4. Call AUTH_CHK() // Authentication check.
  Step5. Call DECYP ( ) //decryption process for
generate plain text.
-------------------------------------------------------
```

All the steps are given below in details. Here public key of patient means the patient-id provided by the hospital or clinic.

5.1 Session Key Generation

Session key [17] is a secret key for symmetric encryption which is used for a particular transaction or session and is valid for a short period of time. In our scheme the approximation algorithm particularly subset sum problem is used to create session key. At first the user chooses the range of numbers and these numbers are used as input to the approximation algorithm. A particular output of the subset sum problem is the required session key.

Algorithm 1 (SSKG ())

```
Input: upper limit (U), lower limit(L) and key length.
  Output: session key.
-------------------------------------------------------
  Methods:
  Step1. Set sum, i, n, tot, v as integer and w [] as
integer array.
  Step2. v←user chosen. //v is the length of the key.
  Step3. L← lower limit, U← upper limit, and tot←0,
n← (U-L) +1. // 5 ≤ n ≤ 100.
  Step4. For i=0 to n
  Step5. w [i]← value between U and L.
           end for
  Step6. Set sum← L+(L+1)+(L+2)+...+(L+v). // v is
any integer chosen by user which provides length of
session key.
  Step7. For i=0 to n
  Step8. tot← tot+w[i]
        end for
  Step9. Call subsum(0,0,tot)
  Step10.End
  Sub Procedure:
  sub_sum(      ) : cs, k, r as integer // sub_sum() is a
function and cs, k, r are parameters.
```

```
  Step1. Set i, x[] as integer.
  Step2. x [k]←1
  Step3. if (( cs+w[k])>sum) then
  Step4. For i=0 to less than or equal to k
  Step5. if (x[i]==1) then
  Step6. print w[i] // the values in w[i] represents
session key.
      endif
         end for
       end if
  Step7. if ((cs+w[k] +w[k+1])<=sum)
  Step8. Call subsum(cs+w[k], k+1, r-w[k])
  end if
  Step9. if ((cs+r-w[k])>=sum) and ((cs+w[k+1])<=sum))
  Step10.x[k] ←0
  Step11.Call sub_sum(cs, k+1, r-w[k])
  Step12.end if
  Step13.end Sub Procedure
```

This session key is used to generate cipher text from plain text and vice versa. This established session key provides confidentiality for subsequent communication and for each session the session key will fresh. This key is send to recipient end through a frame structure which is described below.

5.2 *Intermediate Key Generation*

In our scheme we have used the intermediate key for authentication check between two patients and also for encryption process. Here intermediate key is generated by using two public keys of two patients. So the intermediate key may vary for different pair of patients. To generate intermediate key we have used the concept of linear congruence in number theory, three variables linear function, and Pell's formula. The details algorithm is given below.

Algorithm 2 (INTMKG ())

```
Input: Two public keys of two patients and four constants.
  Output: Intermediate key.
```

```
  Methods:
  Step1. Set d, i, len, fval as integer (as global variable).
  Step2. Set imk [], ps[], pr[] as integer array.
  Step3. Set fval ← key_fn( ). {/* the key_fn( ) is a module which
provides a value
```

in key generation .∗/}
Step4. pr []←public key of receiver.
Step5. ps[]← public key of sender.
Step6. len← get_length (receiver's or sender's public key).
Step7. For i=0 to n
Step8. ps[i] ← ps[i] XOR d and pr[i] ← pr[i] XOR d. /∗ d is a random value between the lowest ascii value of sender and highest ascii value of receiver.
Step9. if (gcd(ps[i], fval)= = 1) then
Step10. imk[i]← call linear_congr (ps[i], pr[i],fval) {/∗ linear congruence, the equation ps[i]x≡pr[i] (mod fval) provides unique solution (Theorem1) which is stored in imk[] . This is the required intermediate key.∗/}
Step11. else imk[i] ← call linear_congr (ps[i]+1,pr[i],fval)
Step12. End if
Step13. End for
Step14. Stop

Sub Procedure:

key_fn ( ): public key of sender, public key of receiver.
Step1. Set $n_1, n_2, n_3 n_4$, d_1, d_2 as integer variable.
Step2. d_1 ← Value chosen by user and d_2← value chosen by user.
Step3. n_1& n_2← randomChar (public key of sender) such that (n_1~n_2) ≥d_1. If it is not possible to get such numbers n_1 and n_2 with said condition then we take n_1 and n_2 such that (n_1~n_2) nearest to d_1.
Step4. n_3& n_4← randomChar (public key of receiver) such that (n_3 ~n_4) ≥d_2. If it is not possible to get such umbers n_1 and n_2 with said condition then we take n_1 and n_2 such that (n_1~n_2) nearest to d_2.
Step5. We get two numbers for each n_1, n_2, n_3, n_4 respectively using the pell's equation (x^2-ny^2=1). Thus we get 8 numbers in total.
Step6. The modulus operation is done on 8 numbers by a meanOf (d_1 d_2).
Step7. Variable number of times the shuffle operation is done on 8 numbers.
Step8. First three or four numbers are used for constant terms in a three variable

```
  linear function f(x,y,z) and again after shuffle operation
remaining numbers
  are used for values of x, y and z in the function.
This module returns the
  functional value which is used for key generation.
  Step9. End sub procedure.
```

This intermediate key is transmitted to the recipient end through secured channel before encryption. In recipient end this key is used for authentication purpose and encryption purpose.

5.3 *Encryption Process*

Now in our proposed technique encryption is done by using intermediate key and session key. To provide non-linearity in encryption process we use circular left shift (CLS) operation [18] in encryption process. A two variable non-linear function (*fn _ enc* (*session key*, *intermediate key*)) is used where ascii value of each character of session key and intermediate key and their difference are used to calculate functional value. The functional value is used as the number of times the circular left shift occurs. Since, we have used a non-linear function so the output of the function provides a non-linearity in cipher text. This type of double encryption with session key and intermediate key provides extra flavor of robustness in our technique. In both cases XOR [14] operation is executed with a CLS operation at end. An example is given below.

Example: let $fn\_enc(x, y) = \frac{x^2}{y} + y^3$ be a non-linear function.

Where $y =$ ascii value of each character of session key, $x =$ ascii value of each character of intermediate key. Now $m = fn\_enc(x, y)\ \%\ (ascii\_diff)$, where $ascii\_diff =$ the ascii difference between each character of intermediate key and session key. The functional value m is used as the number of times the circular left shift occurs in each character of partial cipher text to generate final cipher text.

The encryption algorithm is given below.

Algorithm 3: (ENCYP ()) `Input: - plain text, intermediate key, session key.`

```
  Output: - encrypted file.
```

```
  Method:
  Step1. Set m as integer, plain_file as plain text file
and cipher_file, cipher_final
  as cipher text file.
  Step2. Set output_file as temporary file.
```

```
  Step3. if (!eof ) then
  Step4. output_file= bit_XOROP (plain_file, session
key)
  Step5. cipher_file = bit_XOROP (output_file,
intermediate key)
  Step6. m= fn_enc (session key, intermediate key) /*
one example is given above.*/
  Step7. Cipher_final= cipher_file<<m. // circular left
shift operation.
  Step8. end if
  Step9. Stop
-------------------------------------------------------
```

5.4 Generation of Authentication Code and Transmission File

After encryption with two keys we create a rigorous frame structure (transmission file) with four attributes such as Header, Cipher text [19], Tail_msg, and Pading using the function AUTH_CHK (). Any hash function is used to generate hash value of intermediate key. The function provides a compact form of text which is ready for transmission to the receiver end.

Algorithm 4: (AUTHEN_TRANS ( )) Input: - intermediate key, session key, cipher text and d_1, d_2 chosen by user.
  Output: - transmission file (trans_file).

```
  Methods:
  Step1. Set imk [ ], ssk [ ], Padding [ ], Header_msg
[ ], Tail_msg[],enc_val[] as integer array.
  Step2. Set trans_file as a file. {/* this file is
transmitted to the receiver end */}
  Step3. imk [ ] ← intermediate key and ssk[ ] ← session
key.
  Step4. Padding [] ← hash_ValOf(imk[]).
  Step5. Header_msg [ ] ← (imk[ ] ≪ ((d1 + d2)/4)) XOR ssk [
].
  Step6. Tail_msg[ ]← enc_val[ ] {/* enc_val[] contains
encrypted value of d1 and d2.*/}
  Step7. trans_file←Call concate_text (Header_msg [
],cipher text,Tail_msg[],Padding[]).
  Step8. Stop.
```

If the key size is 16 byte, then total frame structure for transmission is given below which is created by the function AUTHEN_TRANS ().

HEADER_MSG	CIPHER TEXT	TAIL_MSG	PADDING (32 byte)

The Tail_msg part of transmission file contains encrypted value of two constants (d_1, d_2). This encryption is done by RSA technique. Padding field contains hash value of intermediate key. This padding field is used for authentication [20] purpose. The session key is recovered from header part using CLS operation and the intermediate key.

5.5 *Decryption Phase*

In decryption phase, at first receiver generates intermediate key by calling IMKG () and then check authentication using hash value of intermediate key from padding field. After finishing authentication check, session key is generated from HEADER_MSG of transmission file and then decryption process is started using two keys. The entire decryption process is given below by a compact algorithm.

Algorithm 5: (DECYP ( ))

```
Input: Intermediate key, cipher text.
  Output: plain text.
------------------------------------------------------------
  Methods:
  Step1. Set a, n, i as integer and ssk [ ], imk [ ] as
integer array.
  Step2. Retrieve d1 and d2from TAIL_MSG of transmission
file or frame structure.
  Step3. imk [ ] ← intermediate key.
  Step4. Call AUTHEN_CHK (imk [ ], trans_file).
  Step5. if (true) then
  Step6. ssk [ ] ← get_SessionKey( Header_msg of
trans_file,d1,d2). /*session key generation*/
  Step7. Decryption process is done with two keys which
is reverse of encryption process.
  Step8. else Print Authentication fails.
  end if
  Step9. Stop.
```

---

6 Results and Discussions

In this section, simulation results of the proposed technique are presented. All the programs and calculations are done in a machine with above configuration. In our experiments, several sizes of different types of files are used as plain text (Table 1).

Our result section is divided into four parts: (1) Attacks analysis, (2) Randomness test and entropy test of session key, (3) Analysis of encryption technique, and (4) Comparisons among standard algorithms and some other techniques.

6.1 Different Types of Attacks Analysis

There are different types of attacks are exists to break the security when information transmits through network. Generally, intruders hack the encryption key to generate plain text. There are two general approaches—(a) Brute-force attack [21, 22], (b) Cryptanalysis [23, 24, 26].

Key Space Analysis
The size of key space is the total number of different keys in encryption process. The brute-force attack is impractical in such crypto systems where key space is large. Now we consider a general case where secret key is x bits. There are two keys in our proposed technique, first is intermediate key with size x bits and second is session key whose size is $x/2$ bits. Now for the intermediate key, the key space is 2^x and for session key, the key space is $2^{x/2}$ and total key space is $2^{3x/2}$. Using this large key space we discuss the following things.

Brute-Force Attack
A good encryption technique satisfies the requirements of resisting brute-force attack. In this attack, attacker tries to translate the cipher text into plain text using every possible key. On average, half of all possible keys must be tried to achieve success. In most networking system, algorithms are known to all so in this case, brute-force attack will impossible if the algorithm uses large number of keys. At present the fastest super computer is Tianhe-2 having speeded 33.86 petaflops, i.e., 33.86×10^{15} floating point operations per second. Let us consider each trial requires 2000 FLOPS to complete one check. So number of trials complete per second is 16.93×10^{12}. The number of second in a year is $365 \times 24 \times 60 \times 60 = 3{,}153{,}600$ s.

Table 1 Specifications of H/w and S/w used in the experiment

Computer	Lenovo G80
Processor	Intel® Pentium® CPU B950@210GHz
RAM	2GB
Compiler	Turbo C
Disc drive	SA 9500325AS ATA
Operating system	Windows 7 Ultimate (32 bits)

Table 2 Exhaustive search analysis

Intermediate key size (x bits)	No. of trials in standard algorithms (2^x)	Time required (in years) at 16.93×10^{12} decryption/s in standard algorithms	No. of trails in our proposed technique ($2^{(3x/2)}$)	Time required (years) at 16.93×10^{12} decryption/s in proposed technique
56	2^{56}	0.001349	2^{84}	362,289
64	2^{64}	0.34550	2^{96}	1,483,938
128	2^{128}	6.3734×10^{18}	2^{192}	1.1775×10^{38}
166	2^{166}	1.7519×10^{30}	2^{249}	1.6943×10^{55}
192	2^{192}	1.1756×10^{38}	2^{288}	9.3148×10^{66}
256	2^{256}	2.1687×10^{57}	2^{384}	7.3799×10^{95}
300	2^{300}	3.8153×10^{70}	2^{450}	5.4454×10^{115}

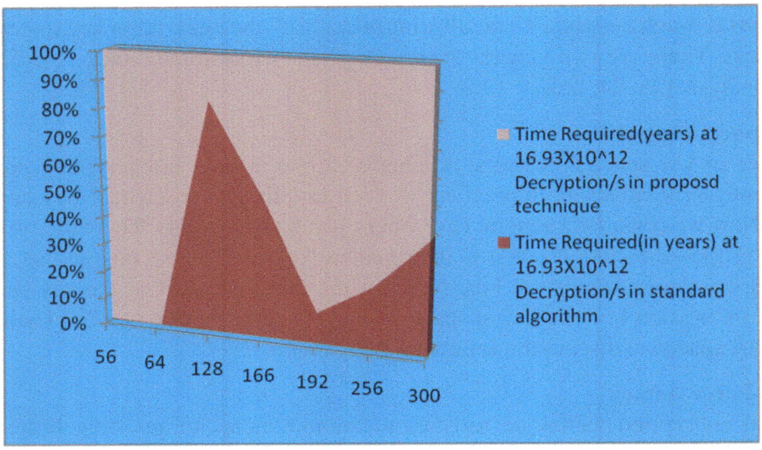

Fig. 1 3D graph of exhaustive key search

Now from the above key space the formula for break the keys is $2^{3x/2}/(16.93 \times 10^{12} \times 3153600) = Y$. So if k increases, then Y increases. The following table and graphs show the average time required for exhaustive key search [8] (Table 2 and Fig. 1).

Observations: From the above table it is seen that with respect to number of trials our proposed technique provides good result than any standard algorithms (like DES, Triple DES, AES, etc.) with same key size. The x-axis of the graph represents key size in bits. The above table and graph also show that our proposed technique provides good result for decryption than any standard algorithms (like DES, Triple DES, AES, etc.) with fixed decryption rate. So, it is difficult for attacker to decrypt any cipher text using assumed key. Thus overall result of our technique is good with respect to any standard algorithms in brute-force attack.

Cryptanalysis Impersonation attacks: Impersonation means an act of pretending to be another person for purpose of fraud. This type of attack involves message or email that seems to come from a trusted one. If an attacker is intended to get other user's information, then the attacker sends a message or email to another user and forge valid information. If an attacker can successfully forge, then an impersonation attack can be done [24, 25].

It is required a strong security policies and vigilance on communication media to stop impersonation attack. Here we have used intermediate key for message authentication and RSA technique for user authentication. After authentication we can generate session key which is used for encryption. Thus our technique provides security against impersonation attack.

Replay attacks: Replay attack is one type of network attack in which an attacker detects a data transmission and it has delayed or repeated fraudulently. As a result attacker can gain access to a network, gain vital information from other, or complete a duplicate transaction. An attacker may launch a replay attack to delay or even stop the response to any request message. If a vehicle receives the request sent by the attacker, it means that the attacker has successfully launched the replay attack and the vehicle may be guided wrongly [25, 26].

To defend against replay attack in this article the concept of session key is used and in every session the session key is fresh. Thus when attacker retransmit or has delayed to send message then the previous key may not work.

Tampering attacks: Tampering attack refers to that an attacker tampers other users' information or data. An attacker may launch a tampering attack to smart vehicular system if it intends to change data illegally [26, 27].

In our proposed methodology, we have introduced a rigorous frame structure with multiple attributes. Among the multiple attributes the cipher text is one so it is hard to detect the range of cipher text. Thus the attacker cannot launch tampering attack easily.

6.2 Randomness Test and Entropy Test of Session Key

In our technique the session key is generated from subset sum dynamic problem. Now to test the randomness of session key we use some standard techniques such as frequency test [5], entropy [9].

Frequency test: The frequency test is the most basic test for randomness checking. The purpose of this test is to determine whether the number of 1's and 0's in a sequence is approximately the same as would be expected for a truly random sequence.

Mathematical Structure of the Test *Frequency (n),* where n is the length of bit string.

ξ: the sequence of bits which are generated by RNG or PRNG.

S_{obs}: the absolute value of the sum of the X_i (where$X_i = 2\mathcal{E} - 1$) is the sequence divided by the square root of the length of the sequence.

1. Conversion to ±1: The zeros and ones of the input sequences (ξ) are converted to values of −1 and + 1 and are added together to produce $S_n = X_1 + X_2 + \ldots + X_n$, where $X_i = 2\mathcal{E}_i - 1$.
2. Compute the test static $S_{obs} = ABS(S_n)/\sqrt{n}$.
3. Compute P-value= $erfc(S_{obs}/\sqrt{2})$.
4. If P-value≥0.01 then the conclusion is that the sequence is random and if P-value< 0.01

   then the sequence is not random.

Observations: NIST SP 800–22 specifies that the randomness test must follow the three characteristics such as uniformity, scalability, and consistency (Fig. 2).

In case of uniformity and scalability, the occurrence of a zero or one is equally likely that is the probability of occurrence of zero or one is half. The above table of frequency test result shows uniformity and scalability of our technique.

In case of consistency, we can say that we have used two keys one session key and another is intermediate key. But these two keys are session oriented by nature. Their seed values are changed session to session. Here we have used concept of dynamic problem and linear congruence based on their public keys of vehicles. Thus there is no similarity between previous key and present key. Since the two keys may change time to time then the next output bit in the sequence of key generation should be unpredictable in spite of any knowledge of previous bits in the sequence.

Table 3 Table for Frequency test result

Intermediate key size (bits)	Frequency test result of our technique	Frequency test result of PRNG()
56	4598.256	4272.772
64	4625.147	4310.446
128	4662.584	4336.617
166	4579.128	4270.683
192	4519.967	4270.656
256	4633.422	4323.351
300	4722.322	4423.356

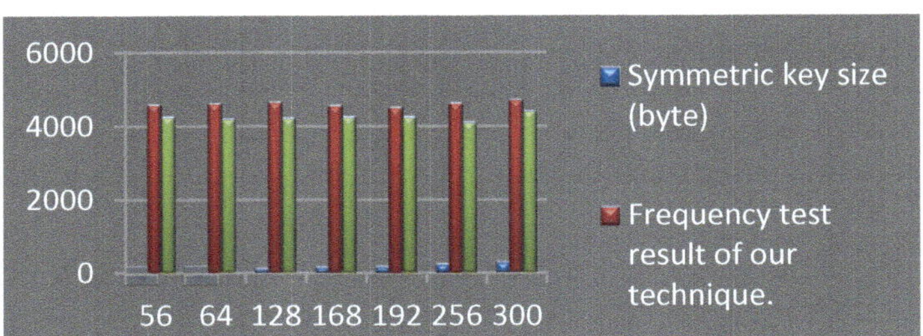

Fig. 2 Graph of frequency test of Table 3

It should not be feasible to determine the intermediate key and session key from the knowledge of any generated values. There is no correlation between intermediate key and generated values. Thus our technique proves the forward and backward unpredictability. Furthermore, from the above table and graph it is seen that our proposed technique provides more randomness than PRNG () which is standard technique.

Experiment on entropy value: Here we describe a comparative study between our technique and standard technique, PRNG () with session key and intermediate key (Table 4 and Fig. 3).

Observations: In cryptography, a cryptosystem is said to be semantically secure if it is computationally impossible for an attacker to extract any information about the plain text from cipher text and its length. Entropy can be defined as randomness or unpredictability of information contained in a message. This randomness breaks the structure of plain text. Entropic security in encryption is similar to semantic security when data have highly entropic distribution [8]. Plain text entropy value is zero. Now from the comparative study of entropy value between our technique and PRNG (), it is seen that the entropy value of our technique is near to the result of PRNG (). The x-axis shows the key length. Thus from the definition of entropic security we say that it is very hard to predict plain text from cipher text if we use our technique to generate session key and the use of this session key and intermediate key in encryption provides robustness.

Table 4 Table for Entropy value

Intermediate key size (bits)	Entropy value of our technique	Entropy value of PRNG()
56	6.88	7.03
64	6.88	7.02
128	6.88	7.02
166	6.88	7.03
192	6.89	7.01
256	6.89	7.03
300	6.90	7.04

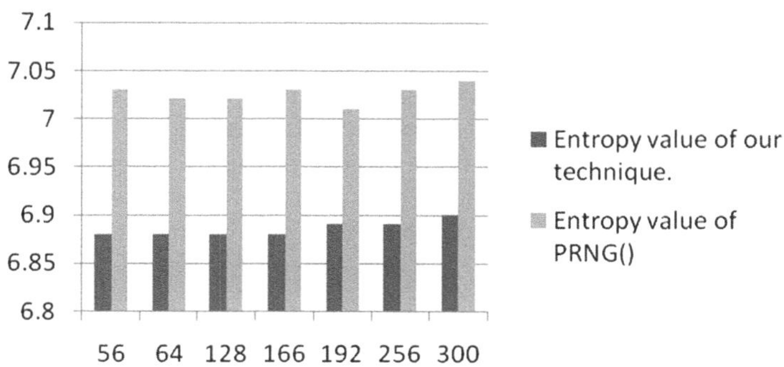

Fig. 3 2D Graph of Entropy value of Table 4

Table 5 Table for Avalanche effect with fixed key

Text size (byte)	Total bit flipped in DES	Total bit flipped in our Technique
17	17	17
24	25	24
32	33	32
64	63	64
92	90	90
128	126	123
144	137	141

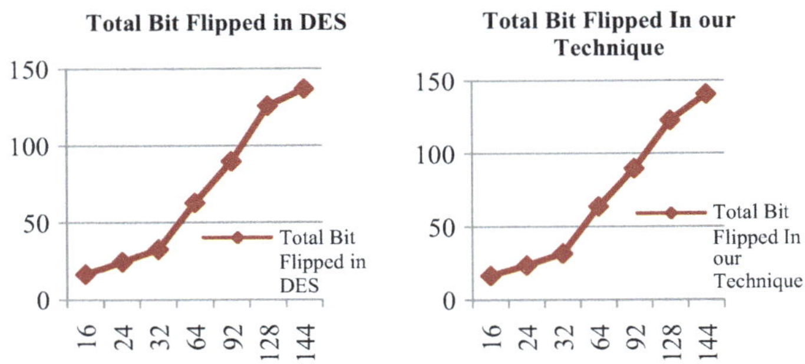

Fig. 4 Graph of Avalanche effect with fixed key for Table 5

Comparative Study on Avalanche Effect

Here a comparative study between our technique and DES on avalanche effect [4, 8] with fixed key is described below with a table and graph (Table 5 and Fig. 4).

Observations: A desirable property of any encryption algorithm is that a small change in either the plain text or the key should produce a significant change in the cipher text. In particular, one bit change in the plain text or one bit in the key should produce a change in many bits of the cipher text. Thus avalanche [8] quantifies the effect on the cipher text when one bit change in plain text. An encryption algorithm that does not provide the avalanche effect can lead to an easy statistical analysis that is if the change of one bit from the input leads to the change of only one bit of the output, then it is easy to guess. Above table and graph provide comparative study between DES and our technique. In the graph *x*-axis represents text size. This study tells that total number of bit flipped in our encryption technique is more than DES. Here we use fixed size key. Thus our technique (using fixed key) provides good result than any standard algorithm (like DES). So, our proposed scheme satisfies the desirable property for encryption algorithm.

6.3 Analysis of our Encryption Technique

Cryptanalysis is the study of cipher text, ciphers, and cryptosystems. The aim of cryptanalysis is to understand how they work and finding and improving the techniques for defeating or weakening them. There are different types of cryptanalysis attacks such as cipher text only attack, chosen plain text attack, known plain text attack, chosen cipher text attack. Breaking an encryption algorithm is basically the finding of the key to access the encrypted data in plain text. For symmetric key encryption, breaking the algorithm usually means trying to determine the key used to encrypt the text. For a public key encryption, breaking the algorithm usually means acquiring the shared secret information between two recipients. The robustness of an encryption technique is depends on non-linearity in cipher text. In our paper we use circular left shift operation and a non-linear function to provide non-linearity in cipher text. As a result our technique is able to protect any types of cryptanalysis attack. The following graph shows the robustness of our protection mechanism.

Observations: Non-linearity is a main theme in any encryption technique. From the above Fig. 5 it is seen that our technique provides more non-linearity in cipher text than simple XOR operation. If we consider any point (x, y) and (a, b) in the plaintext and Cipher_ Proposed, respectively, of above graph, then any periodic gap is not exist between points in the graph, i.e., there is no relationship between two graphs. So it is hard to guess plain text or encryption key from cipher text. Thus our encryption scheme is robust as well as it may protect any types of cryptanalysis like known plain text attack, chosen cipher text attack, etc. From above Theorem 1, it is also seen that this technique satisfies the condition of perfect security because cipher text and plain text are independent.

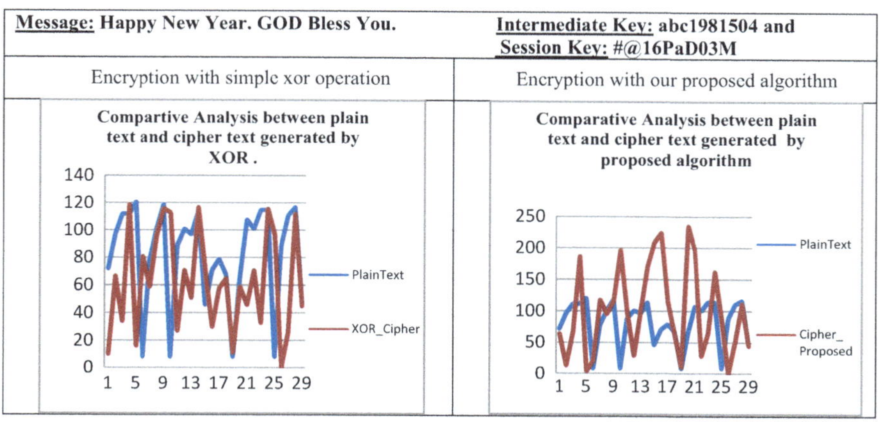

Fig. 5 Analysis of encryption technique

Table 6 Comparisons among proposed technique and standard algorithms

Algorithms	Important features	Important features of our proposed algorithm
IDEA [6, 8]	1. IDEA encrypts 64-bit plain text to 64-bit cipher text blocks, using a 128-bit input key 2. It uses both confusion and diffusion technique 3. A dominant design concept in IDEA is mixing operations from three different algebraic groups of 2^n elements 4. The security of IDEA currently seems that it is bounded only by the weaknesses arising from the relatively small (compared to its key length) block length of 64 bits	1. Our technique encrypts n-bit plain text to n-bit cipher text, using m-bit input key 2. The main design concept of our technique: (a) intermediate key generation using the concept of linear congruence, (b) session key generation using approximation algorithm, (c) circular left shift is used to produce non-linearity in encryption process
RC5 [5, 8]	1. The RC5 block cipher has a word-oriented architecture for variable word sizes $w = 16$, 32, or 64 bits 2. The number of rounds r and the key byte-length b are variable 3. For encryption, there are two steps in each round: (a) bit-wise XOR operation, (b) circular left shift. (c) Addition with the next sub key	1. Our proposed scheme is stream cipher based. Here two keys are used for encryption/decryption 2. Key length is variable 3. For encryption process, there are three steps: (a) Bit-wise XOR operation (b) Non-linearity via non-linear function and (c) Circular Left Shift
BLOWFISH [5, 8]	1. This technique is based on stream cipher. It uses addition, XOR operation for encryption 2. It has a variable key length up to a maximum of 448 bits long which ensures security 3. Blowfish suits applications where the key remains constant for a long time and it is not suitable for packet switching	1. Our scheme is also based on steam cipher. It uses XOR, CLS operations to impose more non-linearity in cipher text 2. The use of double keys and one of this key is changeable by nature which provides robustness in our technique 3. Suitable for packet switching
DES [6, 20]	1. Linear cryptanalysis provides the most powerful attack on DES to date where enormous number of known plain text pairs is feasible 2. Differential cryptanalysis is one of the most general cryptanalytic tools to date against modern iterated block ciphers, including DES. It is primarily a chosen-plain text attack 3. Storage complexity, both linear and differential cryptanalysis requires only negligible storage 4 Due to its short key size, the DES algorithm is now considered insecure and should not be used However, a strengthened version of DES called triple-DES is used	1. Our algorithm is based on stream cipher with two keys one is session key which is changeable in nature. So it protects linear cryptanalysis as well as differential cryptanalysis 2. Our algorithm takes negligible storage for linear and differential cryptanalysis 3. Our algorithm is secure with respect to key size, because we have used two keys with variable length

(continued)

Table 6 (continued)

Algorithms	Important features	Important features of our proposed algorithm
Triple-DES with three keys [8, 20]	1. Triple-DES counters to the meet-in-the-middle attack by using three stages of encryption with three keys 2. Tuchman proposed a triple encryption method that uses only two keys. The function follows an encrypt-decrypt-encrypt sequence $C = E(K_1, D(K_2, E(K_1, P)))$ and $P = D(K_1, E(K_2, D(K_1, C)))$ 3. There is no cryptographic significance in second stage decryption. Its only advantage is that it allows users to decrypt information encrypted by users of single DES 4. There is no practical cryptanalytic attack on 3DES. This method is an improvement over the chosen plain text approach but requires more effort. This attack is based on observation that if the value of first phase encryption and final cipher text is known then the problem reduces to double DES	1. This article proposed a method that counters meet-in–the–middle attack by using two different keys with variable times of CLS 2. This technique has cryptographic importance in wireless network; it uses two keys such as intermediate key and session key which is generated using fuzzy logic. It also provides strong authentication mechanisms 3. It is very hard to anticipate the two keys if plain text-cipher text pair is known

6.4 *Comparative discussion of our proposed technique*

In this section the functionality of our scheme is done by comparing our proposed technology with different standard cryptographic algorithms, different existing schemes [9, 17, 20–23].

Tables 6 and 7 show comparison among different standard algorithms and our proposed algorithm.

Table 7 provides the pertinence of our methodology compared to other existing techniques.

7 Concluding Remarks

At present V2V communication in IoV or any other vehicular communication network is under threat. Therefore we have presented an authentication technique and an encryption technique based on intermediate key and session key for secure data transmission in wireless network for smart vehicular system or IoV. Here receiver decrypts the cipher text using his or her intermediate key and session key. Our technique also provides the proof of identities which enrich the robustness as well as beauty of our proposed scheme. Thus our proposed technique provides two

Table 7 Comparison among some existing schemes and our proposed scheme

Schemes→ Security properties↓	Ref. [11]	Ref. [1]	Ref. [13]	Ref. [14]	Ref. [15]	Ref. [16]	Proposed technique
Confidentiality	Yes	Yes	Yes	Yes	Yes	Yes	Yes
Integrity	No	No	No	No	No	No	Yes
Authenticity (message authentication)	Yes	Yes	Yes	Yes	Yes	Yes	Yes (using intermediate key and RSA)
Freshness protection	No	No	No	No	No	No	Yes
Privacy protection	Yes	Yes	Yes	Yes	Yes	Yes	Yes
Defend against man-in –middle attack	No	No	No	No	No	No	Yes
Defending against tampering attack	Yes	Yes	No	Yes	No	Yes	Yes
Defending against replay attack	Yes	Yes	Yes	No	No	Yes	Yes
Vulnerability	Yes	Yes	No	No	No	Yes	Yes
Defend against impersonation attack	Yes	Yes	Yes	Yes	Yes	Yes	Yes
Cryptanalysis (linear and differential)	Yes	No	No	No	No	No	Yes
Session key establishment	No	No	No	No	No	No	Yes
Defend against information-leakage attack	No	No	Yes	Yes	No	Yes	Yes
Computational complexity and time complexity	Medium	Medium	Medium	Medium	Medium	Medium	Less

contributions on smart vehicular application. Firstly, a new approach for extracting session key and intermediate key uses the concept of dynamic problem and linear congruence method. Secondly, the encryption technique with two keys, i.e., two times encryption with session key and intermediate key and circular left shift operation on partial encrypted form provides robustness of our technique. At last we have included an authentication mechanism in our technique. Comparative statistical tests like entropy value and frequency test, etc. between proposed technique and standard technique prove the sturdier of our key. Lastly, exhaustive key search analysis shows the acceptability of our technique. To the best of our knowledge our proposed technique is the simplest technique with intermediate key, session key with authentication mechanisms. Our proposed scheme can be used in any smart vehicular application or IoV for vehicular network security. It is practically having minimal computational overhead during authentication, encryption, and decryption.

Conflict of Interest The authors declare that they have no conflict of interest.

References

1. Yong, X., Libing, W., Yubo, Z., & Jian, S. (2016). Efficient and secure authentication scheme with conditional privacy-preserving for VANETs. *Chinese Journal of Electron Devices, 25*(5), 950–956.
2. Kahate, A. (2008). *Cryptography and network security*. New Delhi: Tata McGraw-Hill.
3. Hao, Y., Tang, J., & Cheng, Y. (2013). Secure cooperative data downloading in vehicular ad hoc networks. *IEEE Journal on Selected Areas in Communications, 31*(9), 523–537.
4. Buchmann, J. (2004). *Introduction to cryptography* (2nd ed.). Springer.
5. Stinson, D. (2006). *Cryptography: Theory and practice* (3rd ed.). Boca Raton, FL: Chapman & Hall/CRC.
6. Das, A., & Veni Madhavan, C. E. (2009). *Public-key cryptography: Theory and practice*. Upper Saddle River, NJ: Pearson Education.
7. Boneh, D., & Franklin, M. (2001). Identity-based encryption from the Weil pairing. In J. Kilian (Ed.), *CRYPTO 2001. LNCS* (Vol. 2139, pp. 213–229). Heidelberg: Springer.
8. Stallings, W. (2003). *Cryptography and network security: Principles and practice* (3rd ed.). Prentice Hall.
9. Agrawal, A., Gorbunov, S., Vaikuntanathan, V., & Wee, H. (2013). Functional encryption: New perspectives and lower bounds. In R. Canetti & J. A. Garay (Eds.), *CRYPTO 2013, Part II. LNCS* (Vol. 8043, pp. 500–518). Heidelberg: Springer.
10. Chakravorty, J. G., & Ghosh, P. R. (2018). *Advanced higher algebra*. Kolkata: U.N. Dhur and Sons Private Ltd. ISBN 978-3-80673-67-7.
11. Jcobson, M. J., & Williams, H. C., Jr. (2009). *Solving the Pell equation*. New York: Springer. ISBN-13: 978-0387849225.
12. Vijayakumar, P., Azees, M., Kannan, A., & Jegatha Deborah, L. (2016). Dual authentication and key management techniques for secure data transmission in vehicular ad hoc networks. *IEEE Transactions on Intelligent Transportation Systems, 17*(4), 1015–1028.
13. Chuang, M.-C., & Chen, M. C. (2014). An anonymous multi-server authenticated key agreement scheme based on trust com putting using smart cards and biometrics. *Expert Systems with Applications, 41*(4), 1411–1418.
14. Kumari, S., Karuppiah, M., Li, X., Wu, F., Das, A. K., & Odelu, V. (2016). An enhanced and secure trust-extended authentication mechanism for vehicular ad-hoc networks. *Security and Communication Networks, 9*(17), 4255–4271.

15. Li, J., Lu, H., & Guizani, M. (2015). ACPN: A novel authentication framework with conditional privacy-preservation and non-repudiation for VANETs. *IEEE Transactions on Parallel and Distributed Systems, 26*(4), 938–948.
16. Wang, F., Xu, Y., Zhang, H., Zhang, Y., & Zhu, L. (2016). 2FLIP: A two-factor lightweight privacy-preserving authentication scheme for VANET. *IEEE Transactions on Vehicular Technology, 65*(2), 896–911.
17. Joshi, C., & Singh, U. K. (2015). A review on taxonomies of attacks and vulnerability in computer and network system. *International Journal of Advanced Research in Computer Science and Software Engineering (IJRCSSE), 5*(1), 742–747.
18. Ghansela, S. (2013). Network security: Attacks, tools and techniques. *IJARCSSE, 3*(6), 1–11.
19. Oladipupo, E. T., & Alade, O. A. (2014). An approach to improve data security using modified XOR encryption algorithm. *International Journal of Core Research in Communication Engineering, 1*(2), 1834–1838.
20. Delfs, H., & Knebl, H. (2002). *Introduction to cryptography: Principles and applications*. Berlin: Springer.
21. Kak, A. (2015). *Lecture notes on computer and network security*. Purdue University [Online]. Retrieved from https://engineering.purdue.edu/kak/compsec/Lectures.html
22. Zaidan, B., Zaidan, A., Al-Frajat, A., & Jalab, H. (2010). On the differences between hiding information and cryptography techniques: An overview. *Journal of Applied Sciences, 10*, 1650–1655.
23. Chaudhry, S. A., et al. (2017). An improved and provably secure privacy preserving authentication protocol for SIP. *Peer-to-Peer Networking and Applications., 10*(1), 1–15.
24. Gaigole, M. S., & Kalyankar, M. A. (2015). The study of network security with its penetrating attacks and possible security mechanisms. *International Journal of Computer Science and Mobile Computing, 4*(5), 728–735.
25. Yaling, H. (2013). Research on computer network security analysis model. *Computer CD Software and Applications, 2013*(z), 1–152.
26. Ye, H., Li, G. Y., & Juang, B.-H. (2018). Deep reinforcement learning for resource allocation in V2V communication. In *Proceedings of IEEE ICC*. pp. 1–5.
27. Pawer, M. V., & Anuradha, J. (2015). Network security and types of attacks in network. *Procedia Computer Science, 48*, 503–506. https://doi.org/10.1016/j.procs.2015.04.126.

Electric Power-Train Pre-fault Detection Using AI with IoT

Bapu Dada Kokare, Anil Kumar Gupta, and Sanjay A. Deokar

1 Introduction

MG Hector the internet vehicle features:

iSmart innovation gives programming backing to MG hector vehicle [2]. Operating system is android based with arrangement of 4G availability, understands voice algorithm with Indian English pronunciation, vehicle status like everything practically equivalent to world guide to computerized world. In the emergency event that provided airbags are activated, then system triggers an E-Call and an instant message will be sent to the company as well to medical service with the location of the vehicle.

1.1 Crisis Arrangement

The airbags are activated in a mishap, an E-Call will be empowered and an instant message will be sent to the Pulse Hub with the area of the vehicle. The assigned crisis contact will likewise get a crisis call. The Pulse Hub additionally makes a call to the vehicle's head unit and the proprietor's telephone is called after crisis contact if there is no reaction. The Pulse Hub can trigger a rise.

B. D. Kokare (✉)
Department of Technology, Pune University, Pune, India

A. K. Gupta
C-DAC R&D, Pune, India
e-mail: anilkgupt@ieee.org

S. A. Deokar
Someshwar college of engineering, Someshwar Nagar, Pune

© Springer Nature Switzerland AG 2021

N. Gupta et al. (eds.), *Internet of Vehicles and its Applications in Autonomous Driving*, Unmanned System Technologies, https://doi.org/10.1007/978-3-030-46335-9_11

1.2 Crisis Reaction

If associated with a mishap, the airbags are conveyed, an E-Call is enacted and an instant message sent to the Pulse Hub with the vehicle's area. A crisis message is additionally sent to the enrolled crisis contact. The Pulse Hub additionally makes a call to the head unit of the vehicle and if there should be an occurrence of no answer, the proprietor's telephone is called trailed by crisis contact. The Pulse Hub can likewise start a crisis reaction to the specialists whenever required.

1.3 Remote Highlights

Vehicle features are remote cooling control, entryway locks, sunroof, and entryway keys. For example, the algorithms accompanies an Emergency Call highlight that associates with a day in and day out Pulse Hub.

2 Artificial Intelligence [AI]

Artificial intelligence (AI) is an area of computer science is the creation of intelligent machines that work and react like humans. AI always centered around the advancement of PC programs that retrieve information from sensor and use it to learn for themselves.

Deep Learning sets aside more effort to prepare when contrasted with Machine Learning. The principal reason is that there are such a significant number of parameters in a Deep Learning calculation. While Machine Learning sets aside considerably less effort to prepare, extending from a couple of moments to a couple of hours (Fig. 1).

Driving behavior detection using artificial intelligence [4]: By applying knowledge gained from time-series data. It is very easy to understand the behavior of

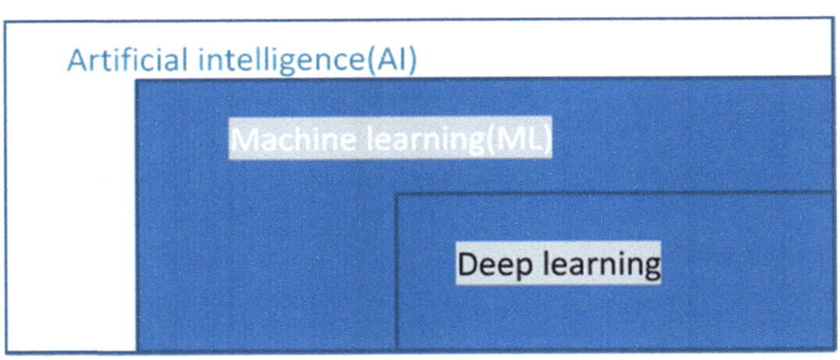

Fig. 1 AI sub-field

driver. For example, driver applies force on pedal and brake of vehicle at the same time then its abnormal condition easily detects abnormal condition that damages the transmission system very badly.

2.1 *AI Application in Gear Transmission System Health Monitoring*

The power-train gearbox information contains vibration [1], gearbox speed rotation per minute (RPM), and temperature of a motor recorded for 30 day at the same time data fed to AI algorithm or assessed with AI methods. This makes it conceivable to recognize an error that is getting set up right on time. A vehicle can be sent to a repair shop immediately, or an algorithm can be intended to record mistake and advance detection in ahead of time. Experimental setup for electric power-train system using Raspberry-Pi.

2.2 *Data Logger Architecture*

Data logger is designed using micro-controller with Ethernet controller. All sensor information logged to local connected server via Mod-bus communication. Data logger build for the use of monitoring driver behavior, vehicle speed, vehicle loading, steering angle, acceleration, braking behavior. Corresponding all channel configured using C programming. Figure 2 shows the drive input logged data of vehicle. From the recorded data it is possible to uncover all required information and understand the behavior of driver. For example, driver changing gear in full speed, algorithm can detect and alert system that cause harmful to transmission system over long time period operation may cause permanent damage to system and reduce the life of transmission system [4] (Fig. 3).

brk detect= No	clutch detect= No	hand break detect= No	seat belt detect= No	seat accupacy= No	accelration detect= No
brk detect= No	clutch detect= No	hand break detect= No	seat belt detect= No	seat accupacy= No	accelration detect= No
brk detect= No	clutch detect= No	hand break detect= No	seat belt detect= No	seat accupacy= No	accelration detect= No
brk detect= No	clutch detect= No	hand break detect= No	seat belt detect= No	seat accupacy= No	accelration detect= No
brk detect= No	clutch detect= No	hand break detect= No	seat belt detect= No	seat accupacy= No	accelration detect= No
brk detect= No	clutch detect= No	hand break detect= No	seat belt detect= No	seat accupacy= No	accelration detect= No
brk detect= No	clutch detect= No	hand break detect= No	seat belt detect= No	seat accupacy= No	accelration detect= No
brk detect= No	clutch detect= No	hand break detect= No	seat belt detect= No	seat accupacy= No	accelration detect= No
brk detect= No	clutch detect= No	hand break detect= No	seat belt detect= No	seat accupacy= No	accelration detect= No
brk detect= No	clutch detect= No	hand break detect= No	seat belt detect= No	seat accupacy= No	accelration detect= No
brk detect= No	clutch detect= No	hand break detect= No	seat belt detect= No	seat accupacy= No	accelration detect= No
brk detect= No	clutch detect= No	hand break detect= No	seat belt detect= No	seat accupacy= No	accelration detect= No
brk detect= No	clutch detect= No	hand break detect= No	seat belt detect= No	seat accupacy= No	accelration detect= No
brk detect= No	clutch detect= No	hand break detect= No	seat belt detect= No	seat accupacy= No	accelration detect= No

Fig. 2 Data logger data view in CSV file format

Fig. 3 Data logger

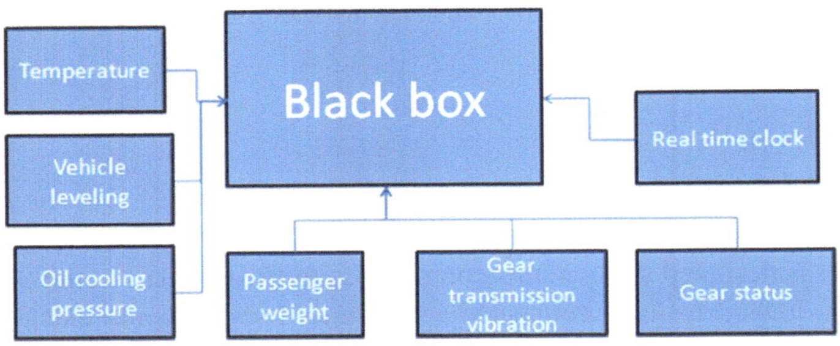

Fig. 4 Black box architecture

3 Black Box Requirements and Improvement

From manually driving to autonomous driving there many question arises like How to understand the machine work nature? How to make decision to drive all possible direction? How to take decision? How to use sensor and remove manual interpretation? How to deploy algorithm to controller? In a mishap, how might you understand what turned out badly without black box?

Application of AI creates many queries and helps to consider all input information while designing black box. I have been designed a data logger which monitor driver behavior contains information of driving behavior, and condition monitoring using vibration sensor, temperature sensor, oil cooling pressure of electric powertrain transmission system.

Vibration sensor records information and monitored online by AI algorithm. It is observed that gear shifting deterioration starts due to wear and tea process at contact of point. Corresponding improvement done in gear manufacturing process regarding hardness. For good results in Electric Power-train performance time to time maintenance is very important (Figs. 4 and 5).

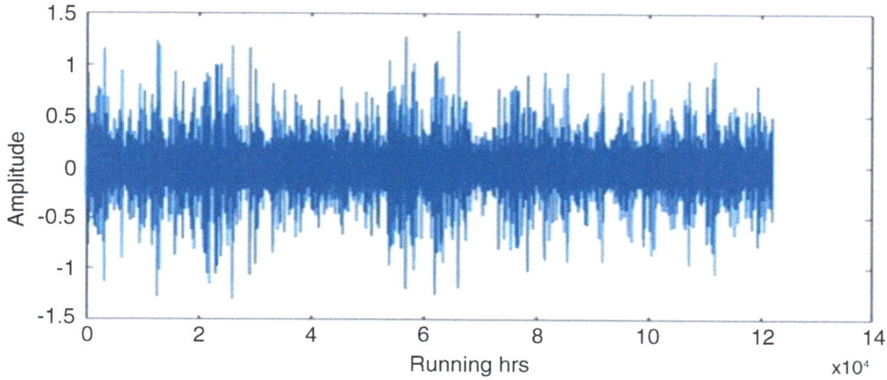

Fig. 5 Recorded data

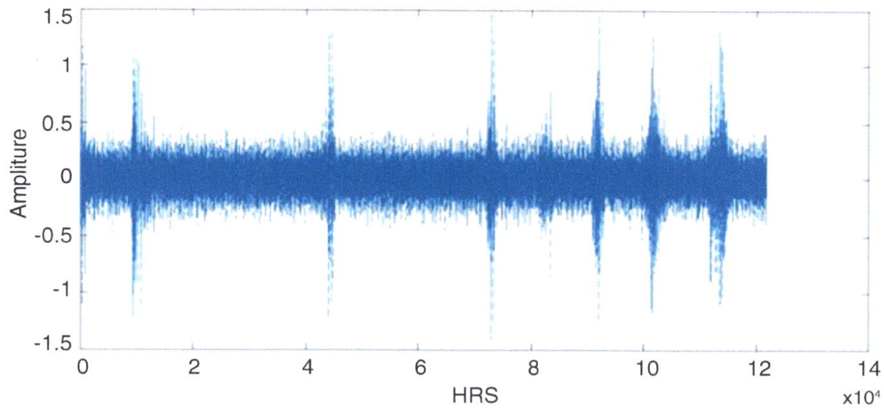

Fig. 6 Recorded data vs peak change value in vibration for 30 days

Gearbox Vibration of Electric Power-Train [3] After viewing the recorded time-series data of a transmission system vibration go on increasing frequently when gear change. Refer Fig. 6.

4 Implement AI Algorithm to Detect Power-Train Fault

Vibration sensor recorded data in black box need to pre-processed to train algorithm in electric power-train gearbox.

Data analysis step by step:

The recorded data is like heavy truck that should be moved to local database for learning and training with the likely behavior of fault data set. For no relevant data tried to reduce specific data set and reduced complexity of the data set. Using large data set is good mindset but it should be processed in the right way.

The following steps are followed to uncover the hidden layer in graph.

Step 1: Define Problem Statement
Formulated with known recorded parameter that data set will help very easy. Like vehicle engine power-train oil cooling pressure and temperature are correlated. If power-train cooling pressure goes down engine cooling will affect mean temperature [3] and increases to high. For this need to collect data regarding temperature, vibration of machine. As shown plotted graph the data set collection.

Step 2: Establish the Connection for Data Recording
I have shown example to record data for using Arduino and Ethernet shield H/W and data logged in .csv format.

Step 3: Set the Minima Maxima with Tolerance
That recorded data feed to algorithm in continuous mode. For occurrence, data range is 0.0–0.5, range should not beyond like.

Step 4: Reduce the Data Set
Analyze the data set which help to save time for begins, or with large data set, expert should not lead to wrong decision. For this there are two methods used to reduce data size (a) top specimen and (b) grouping. Later these methods help to understand sample model result yields expected result or not.

Step 5: Feature Selection [4]
While formulating the problem need to take decision like what type of algorithm should be used. As shown in image AI classification. Those types are supervised learning and unsupervised (Fig. 7).

The classification done with regard to expected answer. For time-series data prediction is possible when previous old value is known. Then take decision from

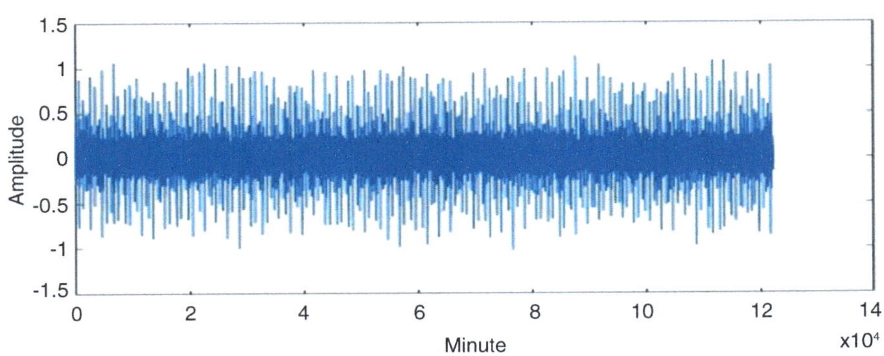

Fig. 7 Recorded data with changing peak vibration

the available graph that driver is driving properly or not. Is there operating condition of worsening transmission?

Cluster Find answer with classes and no. of rules. The difference between cluster and classification is a no. of various group which will part of the problem statement. For this need to select suitable cluster with characteristic information.

Classification Group fault characterization finished with essential data likely kinds of machine, type of vehicle, type of transmission. This particular shortcoming information gathered with vectored vibration calculated and determined resultant vibration signal. Black box gives result in binary format yes or no. Data cleaning is also an important part of feature selection. Data cleaning of non relevant information is also important step of your model designing.

Regression It estimate yield the numeric value. It finds the difference in actual result and expected result.

Ranking In this method rank decide the importance of result and organize answer structurally. That is, Google search result shows most relevant website first with input search query. Feature selection helps to reduce the overall error or over-fitting, improves accuracy, and reduces training time.

Step 6: Data Scaling
It refers to change data variation likely to move averaging reduce min/max value and plot only peak values that have significant values. Also improves the shape of distribution achieved with change in mean value of each feature and scale-up or down by it with non-constant feature by their standard deviation. The scaling functions are normalization, discontinuation of data, categorical selection.

Bearing Sustainability Bearing manufacturing company must pass license of the stress test. Energy consumption proportional to energy transfer transmission.

5 Analysis of Gear Transmission Fault Detection

Versatile Learning Recorded information feed to calculation by on the web. This information gathered from mechanical information lumberjacks' powers to utilization of information science and AI systems. With realize information design progressively and past time information with rehashed situation makes pushing ahead rule. It implies learning machine after some time, if information fits out of the rehashed example framework differentiate the mistake and alert the utilization called versatile algorithm (Figs. 8 and 9).

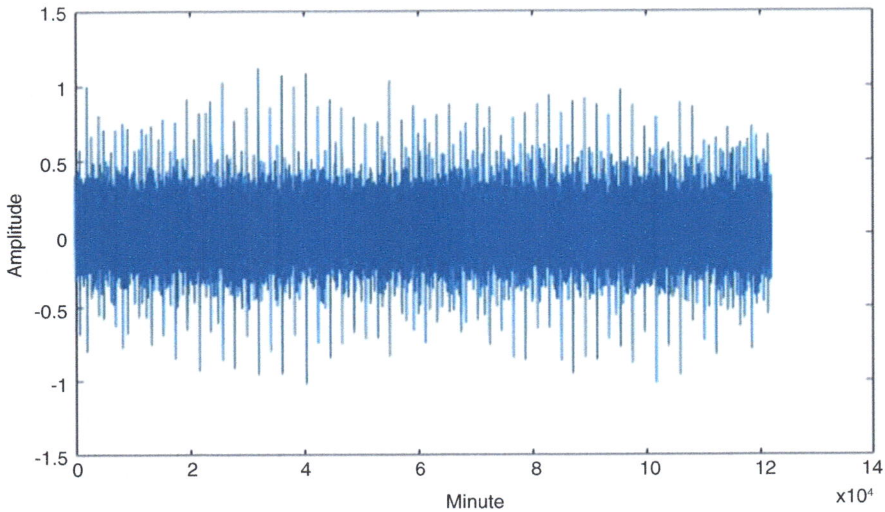

Fig. 8 Moving average data value

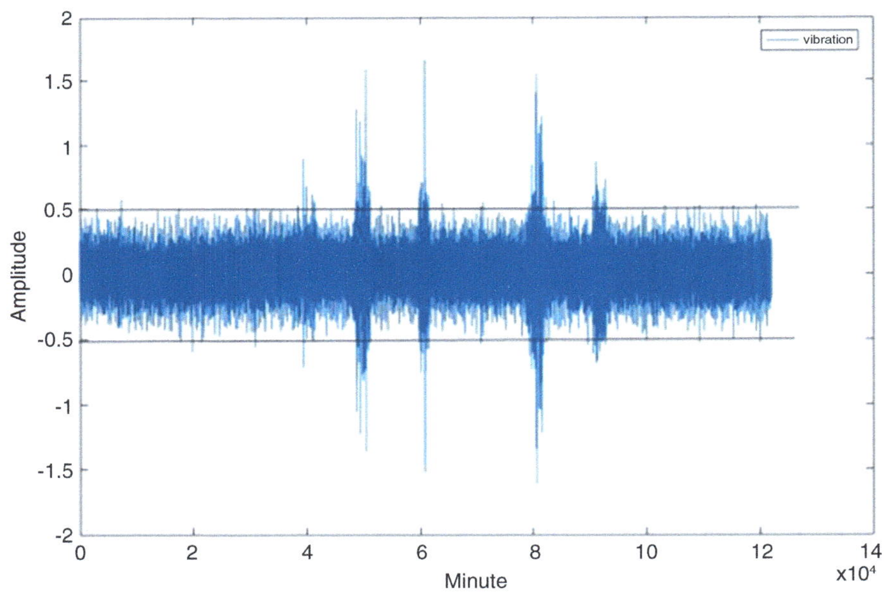

Fig. 9 Gearbox abnormal noise

6 Controller Decision Support System

Neural Networks consist a mathematical representation that mimics the behavior similar to human thought and connects smallest computing units called neurons. One of the main properties of AI object is to be adaptive, i.e. learn and analyze variable conditions, and be able to respond to new, previously known condition that is intelligence.

7 Results and Discussion

Strong fault hides the weak fault. Detecting weak fault for short time behavior are most important parameter to control behavior of vehicle driver. Also experimental setup explained and how to deal with logged data detailed well. For complete failure detection online parameters recorded over good and bad time of machine [5]. To implement AI algorithm for pre-fault and post-fault detection, parameters needed to train and compare record ensures long run powertrain operation and easy to calculate remaining useful life of powertrain over gained experience..

Acknowledgments I would like to express my special thanks of gratitude to my guide Dr.Anil Gupta (C-DAC Pune) as who gave me the golden opportunity to do this wonderful project on the important topic of electric power-train error detection, which also helped me in doing a lot of research and I came to know about so many new things. I am also really thankful to Sanjay Deokar sir for designing controller decision support system

References

1. Liu, J., & Peng, H. (2010). A systematic design approach for two planetary gear split hybrid vehicles. *Vehicle System Dynamics, 48*(11), 1395–1412.
2. Conlon, B. M., Blohm, T., Harpster, M., Holmes, A., Palardy, M., Tarnowsky, S., et al. (2015). The next generation Voltec extended range EV propulsion system. *saE International Journal of Alternative Powertrains, 4*(2), 248–259.
3. Bai, S., Maguire, J., & Peng, H. (2013). *Dynamic analysis and control system design of automatic transmissions* (1st ed.). Warrendale, PA: SAE International.
4. Wallenberg B. F. (1985). Feasibility study for an energy management system intelligent alarm process. In *IEEE Conference Proceedings on Power Industry Computer Applications* (pp. 249–254).
5. Ruddle, A. R., Galarza, A., Sedano, B., Unanue, I., Ibarra, I., & Low, L. (2013). Safety and failure analysis of electrical powertrain for fully electric vehicles and the development of a prognostic health monitoring system. In *IET Hybrid and Electric Vehicles Conference 2013 (HEVC 2013), London* (pp. 1–6). https://doi.org/10.1049/cp.2013.1911

Index

© Springer Nature Switzerland AG 2021

N. Gupta et al. (eds.), *Internet of Vehicles and its Applications in Autonomous Driving*, Unmanned System Technologies, https://doi.org/10.1007/978-3-030-46335-9

The manufacturer's authorised representative in the EU is Springer Nature Customer Service Centre GmbH, Europaplatz 3, 69115 Heidelberg, Germany. If you have any concerns regarding our products, please contact ProductSafety@springernature.com

Printed and bound by CPI Group (UK) Ltd, Croydon, CR0 4YY
30/09/2026
02225498-0004